The Best American Science Writing 2007

THE BEST AMERICAN SCIENCE WRITING

EDITORS

2000: *James Gleick*
2001: *Timothy Ferris*
2002: *Matt Ridley*
2003: *Oliver Sacks*
2004: *Dava Sobel*
2005: *Alan Lightman*
2006: *Atul Gawande*

The Best American 2007 Science Writing

EDITOR: Gina Kolata

Series Editor: Jesse Cohen

AN ECCO BOOK

HARPER PERENNIAL

NEW YORK • LONDON • TORONTO • SYDNEY

HARPER ● PERENNIAL

Permissions appear on pages 331–333.

THE BEST AMERICAN SCIENCE WRITING 2007. Compilation copyright © 2007 by HarperCollins Publishers. Introduction copyright © 2007 by Gina Kolata. All rights reserved. Printed in the United States of America. No part of this book may be used or reproduced in any manner whatsoever without written permission except in the case of brief quotations embodied in critical articles and reviews. For information address HarperCollins Publishers, 10 East 53rd Street, New York, NY 10022.

HarperCollins books may be purchased for educational, business, or sales promotional use. For information please write: Special Markets Department, HarperCollins Publishers, 10 East 53rd Street, New York, NY 10022.

FIRST EDITION

Designed by Cassandra J. Pappas

Library of Congress Cataloging-in-Publication Data is available upon request.

ISBN: 978-0-06-134577-7
ISBN-10: 0-06-134577-6

07 08 09 10 11 BVG/RRD 10 9 8 7 6 5 4 3 2 1

Contents

Introduction by Gina Kolata

I'VE WRITTEN THOUSANDS of articles in my career as a science writer, and so when I'm asked which was my favorite, it's hard to reply. A favorite. It would have to be an article that was exciting to write, because the subject was just so arresting, and one that I can read years later and still savor. The problem, if you can call it that, is that there have been so many celebrated, surprising, and unprecedented ideas and discoveries in the past few decades that the list of great subjects is longer than you might expect. And every year, I write something that so absorbs me that it becomes my new love.

Yet despite all those caveats, I actually do have one article that rises to the top of my list. It's one I wrote in 1993 that announced the solution of a longstanding and notorious math problem, Fermat's Last Theorem.

I studied mathematics—I have a master's degree in math—and this was one problem that I never thought would be solved in my lifetime. It had bedeviled the world's greatest mathematicians for 350 years and had obsessed amateur mathematicians who were sure they could find a way to solve it. It seemed so simple, just a prediction about what would happen if the sort of equations in Pythagoras's theorem involved powers greater than 2. Over the years, thousands of false proofs were published. Some famous mathematicians said the problem was just too hard and they were simply not going to waste their lives trying to solve it. Then, seemingly out of nowhere, a quiet Princeton mathematician, Andrew Wiles, announced that he had a proof of Fermat's Last Theorem. He'd spent ten years holed up in an office in the attic of his home, telling no one what he was doing, working at a tiny wooden desk, convinced he had a method that would work. And he did.

In my article, I quoted Kenneth Ribet, a mathematician from the University of California at Berkeley, telling what that proof meant. "The mathematical landscape has changed," Ribet said. "You discover that things that seemed completely impossible are more of a reality. This changes the way you approach problems, what you think is possible."

That remark captured the excitement of the moment. I remember the night I wrote my story—it would appear the next day on page one of the *New York Times*. I came home from work, put on my running clothes, and dashed out the door, running through the darkened streets of my neighborhood. I felt like I was flying, high on the excitement of the discovery, thrilled that I had seen the day when the problem was solved, and overwhelmed that I had written a story that announced that proof to the world.

Stories like the one about the proof of Fermat's Last Theorem do not come often, but I am constantly amazed by how many incredible discoveries and new ideas are advanced every year. And that is what makes articles about science unlike any other field of writing. It is why I became a science writer and why I have become increasingly convinced that science is the greatest journalistic subject of our time. Advances in science have changed who we are as human beings and they are changing what we will become.

That's a bold statement, I know, but think of the many examples. In past centuries, there were the discovery that the earth rotates around the sun, the discovery of evolution, and the discovery of relativity theory, of quantum mechanics, of DNA, all of which led to philosophical and scientific upheavals. Then there were the great medical discoveries—the germ theory of disease, anesthesia, vaccines, antibiotics. They helped change the very trajectory of human history and they are a major reason why our expectations of what our lives can be are so radically different than they were as recently as the nineteenth century. No longer do disease and death stalk us from early in life. Along with the medical advances there was the revolution in agriculture that eliminated starvation in much of the world.

More recently, there was the world-changing emergence of computers and the Internet. I could go on—genetic engineering, cloning, global positioning systems. The list could almost become tedious. But the impact of discoveries in science and technology is breathtaking and the pace of discoveries seems to be accelerating. To take a single example, just a few decades ago, there was no e-mail, no Internet, no word processing, not even a fax machine, and not even an express mail service.

The exhilaration of those discoveries, and their impact, is what science writing can convey. But great science writing also means articles that are written so well that it is a pleasure, not a chore, to read them.

And so, when I was asked to select examples of the best science writing from 2006, I looked for two things—writing that engaged me immediately, and ideas that might change the way we view the world. My greatest difficulty was winnowing the field. In the end, I chose articles that drew me in with startling notions or vivid images or that developed their stories with unusual skill. I looked for what the advertising industry calls the grease spot—a place near the start of an article that made me just slide right in. Every one of the articles in this book has one. Those I chose include stories of medicine and medical ethics, of computer science, linguistics, evolution, global warming, mathematics, and public health. The reporters' writing styles range from the intensely personal to the distant, almost detached. Some stories are told in the first person; some are told with paragraphs-long direct quotes from scientists. The articles range from the provocative to the informative, from the kooky to the hilarious, and they illustrate, to me, that science remains one of the most exciting areas of journalism today.

Among the many articles on medicine and medical ethics, for example, I chose Lawrence K. Altman's gripping *New York Times* story about Dr. Michael E. DeBakey, the celebrated heart surgeon, "The Man on the Table Was 97, But He Devised the Surgery." The inner lining of DeBakey's aorta, the main artery leading from his heart, ripped one day, and the only way to save him was with a demanding and difficult operation, one he himself had developed. But Altman's article was much more than a story of a surgeon whose discovery cured himself. It was about who decides whether to do a risky operation on an elderly man who had said he did not want heroic measures. It was about anesthesiologists who refused to help and a determined wife who insisted on the operation. And it was about how a person's ideas of what they want at the end of life might change when the end is nigh. Altman's article also raised questions about the cost and consequences of heroic measures. DeBakey's recovery was prolonged and rocky and his medical costs were more than a million dollars. I was riveted by the story from the start and could not put it down.

I also liked Jerome Groopman's *New Yorker* article, "Being There." It raises an issue I had never considered, and in an unforgettable way, starting with the story of a Michigan state trooper who was shot in the back and rushed to an emergency room. His wife wanted to be with him but the hospital's policy was that family members could not be present during resuscitations. The trooper's wife managed to persuade the hospital to let her watch the resuscitation efforts. Should they?

A good science writer can also take a topic that has come to seem almost hackneyed and make it fresh and compelling. That is what Elizabeth Kolbert does in her quietly provocative *New Yorker* article, "Butterfly Lessons." It's about the discoveries of how insects and toads are responding to the warming planet. In other hands, a subject like that might have made me yawn. But Kolbert's beautifully written article had a message that I could not forget, told in the words of the scientists themselves. Chris Thomas, who studies migration patterns of butterflies, summed it up: "If there is this overwhelming evidence that species are changing their distributions, we're going to have to expect exactly the same for crops and pests and diseases. Part of it simply is we've got one planet, and we are heading it in a direction in which, quite fundamentally, we don't know what the consequences are going to be."

There is, of course, real scientific debate over the extent of the human contribution to global warming and how much the warming trend would be affected by the sorts of carbon dioxide reductions that are feasible. One of the challenges of science writing is to report on such controversies fairly and without sounding adversarial. This year, there was an article that seemed to me to be an exemplar of the craft of writing about a divisive topic and it happened to be an article on a debate over global warming. It's a *New York Times* article by William J. Broad, "In Ancient Fossils, Seeds of a New Debate on Warming," and it starts out gently by discussing the advances made in understanding ancient climates. But after you slip into the story, you will find yourself noticing not just the science but the politics of this debate. For example, Robert Giegengack of the University of Pennsylvania, a geologist who studies ancient atmospheres, finds no relationship between global temperatures in the past and carbon dioxide levels. He says other scientists have told him to just stop broadcasting that finding, telling Broad, "People come to me and say, 'Stop talking like this, you're hurting the cause.'"

In their *New Yorker* article, "Manifold Destiny," Sylvia Nasar and David Gruber tell the fascinating story of an astonishing mathematical discovery—a proof of the Poincaré conjecture. That conjecture was one of math's most celebrated problems and one that had defied mathematicians' best efforts for a century. The Clay Institute in New York had offered a million dollars to anyone who proved it and it was a sure shot for mathematics' most prestigious award, the Fields Medal. What followed is a gripping story that involves not just the problem itself but the question of what constitutes a proof and who should get credit.

In physics, the most publicized great problem involves string theory, a way of unifying the forces of nature. It is abstract in the extreme, highly mathematical, and highly controversial. It also has been the subject of so much science writing, from books to articles to lengthy book reviews to opinion pieces, that it almost seemed that there was no way to make that subject fresh. But Tyler Cabot's article in *Esquire*, "The Theory of Everything," defied my expectations. The grease spot comes in the first paragraph, when we meet a theoretical physicist, Nima Arkani-Hamed, who is sleepless in anticipation of the completion of a new particle accelerator. In eighteen months, Cabot writes, "someone will flip a switch" and protons, flying "at nearly the speed of light" will "smash together hard, harder than any subatomic particles have ever been smashed together on earth." Cabot continues: "It's the greatest, most anticipated, most expensive experiment in the history of mankind. And if Arkani-Hamed is right, it could help prove that the laws governing the universe at every scale—from the smallest quark to the largest black holes—are one and the same. Or else, of course, it could prove that Arkani-Hamed is full of shit."

Good science writing can also be lighter in tone, of course, but sometimes a light tone can be even harder to pull off than a serious one. So, after all the sturm und drang about the teaching of evolution, I was delighted by "God or Gorilla," an article in *Harper's* magazine by Matthew Chapman, who is the great-great-grandson of Charles Darwin. It begins with a brief paragraph about a lawsuit over teaching evolution in Dover, Pennsylvania. Then, in the next paragraph, Chapman wrote: "That's the basic story, but if you think you know everything there is to know about this, you are wrong. Only I know the truth."

That hooked me, and I was not disappointed. I won't spoil the fun by revealing much more, except to say that Chapman's writing is a constant delight. He does not pretend to be objective, but his sharp eye and vivid words made the story come alive. Here is an example, a description of a scientist who has made his career espousing intelligent design: "Behe put his hands behind his head and leaned back in his chair, smiling defiantly. He looked like a naughty child who had told his mother he'd seen a ghost and wouldn't budge from the story no matter what."

Another entertaining article, Patricia Gadsby's "Cooking for Eggheads," from *Discover*, is irresistible for anyone who, like me, loves to cook. In it, we come to know an energetic Frenchman who is the head of the molecular gastronomy group at the College of France. He focuses on the basics: Can you improve on the 10-minute egg? Well, Gadsby explains, "If you're willing to learn a little egg chemistry, you can calibrate your eggs with astonishing exactitude."

The darker side of science, the questions of fraud and questions of how to advance an unpopular idea, are also represented in the collection. "Truth and Consequences," from *Science* magazine, tells what happens when graduate students suspect their advisor is cheating on his data. "Schweitzer's Dangerous Discovery" discusses a paleontologist whose work was so controversial that one reviewer of Mary Higby Schweitzer's paper told her that her findings just could not be true. "I wrote back and said, 'Well, what data would convince you?' And he said, 'None.'"

If nothing else, the articles show what a human endeavor science is, and, in the hands of these skilled reporters, what memorable tales can be told.

The Best American Science Writing 2007

TYLER CABOT

The Theory of Everything

FROM *ESQUIRE*

String theory. M-theory. Loop quantum gravity. The holographic universe. Several theories are competing for the solution to physics' ultimate problem: finding a single theory to unify all of the forces of nature—a theory of everything. Tyler Cabot looks in on what the contending theories' proponents are thinking as a new particle collider gets ready to test their hypotheses.

EIGHTEEN MONTHS TO GO. And now some nights Nima Arkani-Hamed can't sleep. Because in eighteen months someone will flip a switch in something called the Large Hadron Collider in Switzerland. And when that switch is flipped, billions of protons will fly around a seventeen-mile loop at nearly the speed of light until they smash together hard, harder than any subatomic particles have ever been smashed together on earth. It's the greatest, most anticipated, most expensive experiment in the history of mankind. And if Arkani-Hamed is right, it could help prove that the laws that govern the universe at every scale—from the smallest quarks to the largest black holes—are one and the same. Or else, of course, it could prove that Arkani-Hamed is full of shit.

IT'S A FOOL'S ERRAND, this quest for a theory of everything. And Arkani-Hamed is only the most recent of thousands of theoretical physicists to embark on it. The idea seemed logical enough when Einstein first set out on it in the 1920s. If general relativity explains the universe from afar—why gravity pulls the earth around the sun—and quantum mechanics explains the world up close—how atoms, protons, and neutrons react to electromagnetism and the strong and weak forces—surely there must be a way to put the two theories together. After all, whether cosmic in size or minuscule, the particles and forces that govern our universe were all born at the same primordial moment. Yet Einstein failed. And in the interim, armies of physicists, equipped with similarly well-intentioned yet ultimately faulty or unprovable ideas, have followed him to the same well-trod dead end.

Since the mid–1980s, the leading contender for a grand unifying theory has been string theory. The idea is deceptively simple: At the core of every particle in the universe is a tiny thread of energy. Each of these filaments vibrates like a violin string, and its rate of vibration determines its vital characteristics, or tone. There are neutrino strings and electron strings, photon strings and graviton strings. When played together, they compose the symphony of the universe. Or at least, that's the theory.

There's a problem, though. The strings have too much range. So much, in fact, that for string theory to agree with the established laws of physics and mathematics, there must be not three but at least ten dimensions (including time) that are curled up and tucked away. And because each of these multidimensional landscapes requires a different string tuning, there are potentially billions and billions of different versions of string theory relating to billions and billions of different universes.

Then there's the problem of testing string theory. That's how science works. We hypothesize, then we test. And if a hypothesis passes muster, it becomes law. But the strings that supposedly make up our universe are so infinitesimal—one string is to an atom as a single atom is to the entire solar system—that critics argue that we may never be able to build a collider powerful enough to find them, even the collider that Arkani-Hamed stays up all night thinking about.

So here's the latest tally: Number of years since string theory became dominant: 20. Number of potential string-theory solutions: 10^{500} (the number of atoms in the galaxy squared and then squared again). Number of testable theories: 0. In other words, Arkani-Hamed better be at least partially right, because the natives are getting restless.

IF THE PROBLEM WITH STRING THEORY, as some critics claim, is that it's a closed-minded boys club whose lifetime members hopelessly shuffle and redeal the same deck of equations ad nauseam, then the solution may be found at the Jane Bond, a bar in the staid Canadian college town of Waterloo. The Jane Bond has a decidedly grungy 1970s flair. Tattooed hipsters talk with awed reverence of Brooklyn while DJs spin eclectic and esoteric music next to the bathroom, near the disco ball. And then there are the physicists from the Perimeter Institute for Theoretical Physics who have made the Jane Bond their watering hole. They talk theory sometimes. But mostly they just bullshit. "You want to know the true story?" goads a young postdoctoral researcher at that magical hour in any bar when only bad things can happen. "It's the post-9/11 theocons." Just like the rest of America, he continues, the science establishment is afraid of anything new. It doesn't want to consider any alternatives. "The string theorists just masturbate to their same ideas." At this, the rest of the table—a mixed group of young cosmologists, quantum-information theorists, and quantum-gravity buffs—breaks into nervous laughter. Yes, their friend is drunk. But he's right in a general sense, they concede. There is a growing fissure in the physics world between the haves (string theorists) and the have-nots (everyone else). But not at Perimeter, they caution. Perimeter is different.

The first thing you notice when walking through the concrete-and-glass hallways of PI are the lounges with blackboards. They are ubiquitous. And at each one there are usually two or three young physicists—mostly men, most in their late twenties or thirties—arguing over equations. The feeling is more dorm-room TA session than serious discussion about the origins of the universe. Sneakers and jeans rule. The researchers come and go as they please, and they work as they please. And when they grow too tired of drilling through equations and erasing equations and drilling through them some more, they might take a break. There's a squash court near the billiard table, a few floors below the bistro and bar. But don't get the wrong idea. Foosball aside, the physicists at PI are doing serious work.

Perimeter was founded by Mike Lazaridis, founder and co-CEO of Research in Motion, maker of the BlackBerry. As the story goes, Lazaridis, who went to college in Waterloo, thought the scientific world was much too focused on areas of research that promised immediate results and fast returns on investment. Nobody was willing to fund basic research into arcane fields, like the foundations of quantum mechanics. So in 2000 he cut a check for $66 million and convinced two partners and the Canadian government to chip in tens of millions more. His plan was to build a physics institute that was different, a place where physicists would have the freedom to probe more foundational physical questions. Along with executive director Howard Burton, he envisioned a true community of scholars, where physicists from disparate disciplines would cross-pollinate in a non-competitive environment. It now has sixty-four resident researchers, including ten faculty members.

Taking up residence here is a bit like joining the priesthood. You're segregated from the rest of the world, and your job is to get into God's head and figure out how the big damn machine works. And though you can work with others, often you're alone, stuck with only the equations and pictures in your head. It's like exploring a forest, says Andrei Starinets, a postdoctoral researcher from the former Soviet Union who studies string theory and black-hole physics. You can see the forest ahead, he says; it's tall and lush and filled with swamps. The task is to figure out how to enter, which path will have the least sink-holes and booby traps. So that's what he and his colleagues do all day. They gather in PI as if it were a fort, planning a means of attack, looking for the paths of least resistance. This means twisting and retwisting equations as if the search for a unified theory were the world's biggest game of sudoku.

Here's how Laurent Freidel, a faculty member in quantum gravity and particle theory, describes the search: "You feel that there is a beast running in the woods. And you don't let it go, you don't stop. And sometimes if you need to, you don't stop at all for two or three days in a row. But that's the fun part, when you're on track and when you know something's out there. There are no rules. You need intuition to make a connection. And then you have to gather evidence; the more evidence you gather, the more you know you're on the right track. The key for me is not to let go, to continue until I reach it. And there's always a way."

Last year, when Freidel discovered a possible rigorous mathematical solution for the strong force—which acts as the glue between protons, neutrons, and nuclei, and which to that point had been studied only by approximation—he didn't sleep for two weeks straight. "His wife ran into me," recalls a colleague, "and she said, 'Can you do something? He's going insane.'"

On the third floor of Perimeter, at the far end of the hall, is a small office with a small sofa wedged next to an overflowing bookshelf. And on that couch, dressed more like a New York artist than a theoretical physicist—black, black, and more black—is Lee Smolin. One of PI's initial faculty hires, Smolin, fifty-one, began his career in string theory before becoming fed up with the lack of progress and turning instead to loop quantum gravity, an alternate possible unified theory. Unlike string theory, which critics describe as background dependent—i.e., space and time are constant and unexplained—LQG posits that space, time, and even people are all formed from the same network of interconnected loops and nodes, which take on electrical charges when twisted. If the tension between string and antistring theorist was once a family argument, Smolin, the author recently of *The Trouble with Physics*, is the person who decided to air the dirty laundry.

"This is an experiment," says Smolin of PI. "Like any experiment, it's a risk and it could fail. The most important question is, Will important science get done here? And I think there are already examples of things that happened here that would not have happened elsewhere, because the people would not have been in touch with each other."

Things like safer cryptography, recent advances in loop quantum gravity, and a possible refutation of special relativity, the law that nothing can move faster than the speed of light. But what about the big questions? Was it really necessary to blow up the model of how a science institute should function in order to break the stalemate in theoretical physics and eventually discover a theory of everything? "I think that the pragmatic, antiphilosophical thing played itself out," says Smolin, referring to results-oriented physics.

"I think what's going to succeed in the big-open-questions part of gravity unification and so forth are going to be approaches that take the foundations and the fundamental questions more seriously than they have been. And why do I believe that? Because if these problems could have been solved by this very pragmatic approach, they would have been. Because a lot of really smart, motivated people have been working on these problems for three decades in that frame of mind, and if it were possible to solve these problems, they would have done it. I should say, we would have done it. Because it was my generation."

FAR AWAY FROM WATERLOO AND PERIMETER, both in geography and state of mind, stands an old clock tower surrounded by vast fields of overgrown grass and hickory trees. It was here that Albert Einstein first began pondering a grand unified theory in earnest. And it is here at the Institute for Advanced Study, in Princeton, New Jersey, that some of the world's most prominent string theorists—including their master guru of sorts, Edward Witten—now gather in their own fort. If PI is a model of the theoretical-physics think tank of the future, then IAS—like the storied Ivy League university next door—is a reminder of the well-mannered past. You won't find a bar or a foosball table here. But tea is served daily at 3:00 P.M.

At six two, Witten is big both in size and presence. He began focusing on string theory in the mid–1980s, soon after the first string revolution suggested that it was a viable theory of everything. Ten years, a MacArthur grant, and a Fields Medal in mathematics later, he ushered in the second revolution by postulating that the five main competing string models of the time were all part of a bigger, more complex model, which he termed "M-theory." His synthesis—the name of which is still a matter of fierce debate in the physics world Mother? Membrane? Magic? Masturbation?—broke a logjam that had snared progress in string theory for nearly a decade and added to his mystique as perhaps the true heir to Einstein.

But now string is at another logjam, in which there are literally billions of possible string-theory solutions and perhaps no means of testing any of them. "Well, you can't have your best year every year," he says of the frustration in the field, weighing every word very carefully in a voice that is just a few decibels above a whisper. "I've lived through two periods, the mid-eighties and the mid-nineties, where for about six or seven years, roughly, there were a lot of really interesting results that were also relatively easy. And I've also lived through several periods by now where you have to work a little harder to get something interesting."

To Witten, the game is far from over for string, and he hopes the something interesting will be found in Geneva. When it begins tests about a hundred yards below the border of Switzerland and France in spring 2008, the Large Hadron Collider and its 1,232 thirty-nine-ton superconductor magnets will propel billions of protons with seven times the strength of the current strongest particle collider in Illinois, and will mimic the conditions of the universe a millionth of a millionth of a second after the big bang.

At the very least, Witten believes, the LHC should be able to explain the lack of symmetry between electromagnetism, which shapes many of the phenomena of daily life, and "weak interactions," which affect the decay of subatomic particles and are related to radiation. "There's something deeply, deeply wrong if the LHC doesn't discover that," says Witten. Both interactions appear nearly identical at the atomic level (they are grouped together as "electroweak interactions" in the standard model) yet behave very differently in the real world.

One possible explanation is the Higgs boson, or "God particle," which has never been seen or measured but which theorists speculate could be responsible for giving all particles mass. According to Witten, its discovery would be a simple, long-theorized solution to the problem of electroweak "breaking," yet carries numerous pitfalls of its own. For example, the value of the Higgs mass has only been estimated so far; an actual measurement may well require adjustment of that value, which could carry huge implications for how the whole machine—our universe—is put together and whether other universes tuned with different Higgs bosons might exist.

A different, even more extreme explanation for the symmetry breakdown is known as supersymmetry, which theorizes a set of counterparts to our known subatomic particles that are embedded in the architecture of space-time. Besides explaining electroweak interactions, the discovery of supersymmetric particles, with cool names such as squarks, sleptons, and selectrons, would be a huge boon to string theorists, whose model of the universe depends upon them.

Finally, there are the wild-card explanations, such as very large dimensions or low-scale string theories, or perhaps solutions that physicists have not yet even dreamed of. Each of these would vastly change our entire outlook of the universe and our place in it. Luckily for us all, perhaps, the chance of discovering them any time soon is rather unlikely.

JUST DOWN THE HALL from Witten is another leading theoretical physicist who also speaks in hushed tones. Like his mentor, Witten, twenty years ago, Juan Maldacena, a thirty-eight-year-old Argentine, is regarded as one of the great young thinkers of his time. Need proof? Well, here's the song:

> *You start with the brane*
> *and the brane is B.P.S.*
> *Then you go near the brane*
> *and the space is A.D.S.*
> *Who knows what it means?*
> *I don't, I confess*
> *Ehhhh! Maldacena!*

Sung to the tune of "Macarena," those words were used to serenade Maldacena at the 1998 string-theory conference in Santa Barbara. The occasion was Maldacena's newly published work on black holes, which became known as the Maldacena conjecture. It's complicated stuff, but by demonstrating a relationship between quarks and black holes, Maldacena showed that quantum field theory could be used to solve string equations. This was huge, a method of inquiry that might finally bridge the gap between the forces that govern the cosmos and the forces that govern particles. Nearly overnight, thousands of string theorists got to work on Maldacena's work. And at the conference, it was an occasion to sing and dance.

But there's more. The way Maldacena solved his conjecture—by converting complicated five-dimensional equations into four-dimensional equations, then back again—was a discovery in and of itself, leading to an even more jarring conclusion: Gravity and time could be an illusion. Just like the shimmering holograms we grew up with—say, Michael Jordan jumping out of a shiny silver sticker for a slam dunk—our universe could be a giant hologram: a massive two-dimensional plane encoded with quantum information at the edges that makes it appear three-dimensional. It's a mind fuck, for sure, but the payoff could be huge. Holograms could provide an explanation of how a theory of everything might relate to the whole universe and beyond.

Think of our universe, or dimensional landscape, as a giant DVD floating among an infinite number of other DVDs. Each two-dimensional DVD was built in the same factory according to the same theory of everything, yet each one is embedded with a different movie. While our DVD shows a three-dimensional universe ruled by the standard model, the DVD landscape next door could be embedded with a five-dimensional movie and a separate, slightly different standard model. If we could view all of these separate landscapes in four dimensions—perhaps the equivalent of a universal HD-DVD player—we might be able to glimpse the underlying architecture that they all share.

If Maldacena is right, the holographic principle could reveal the order behind everything; there might be an infinite number of universes, but they'd all be ruled by the same laws, just experienced slightly differently.

Still a mind fuck, but not crazy.

FOR CRAZY YOU HAVE TO GO about 250 miles north. At thirty-four, Nima Arkani-Hamed, born in Houston to Iranian physicists, is in the sweet spot of his career: young enough to still have fresh, insubordinate ideas; old enough to have the wherewithal and grounding to push fresh and insubordinate ideas. His office at Harvard is clean and minimal, all polished wood offset by a floor-to-ceiling, wall-to-wall blackboard. He's made a career of producing models of the universe that are staggeringly elegant. And so it figures that when he speaks—and he speaks a lot—it is with authority and simplicity. Plus there is his overwhelming sense of urgency, excitement, and swagger—forget the antistring polemicists! They're just reactionaries! This could be the greatest discovery of our time!

Nima, as everyone calls him, first stunned the physics community in 1998 by postulating—along with Savas Dimopoulos and Gia Dvali—that unknown extra dimensions could be far larger than anyone ever thought possible, perhaps even nearly a millimeter wide. This was counter to everything physicists had theorized about hidden dimensions, which were believed to be only one hundredth of a thousandth of a trillionth of a centimeter wide. In an insular community where hidden universes, parallel realities, and black holes are discussed with the nonchalance of the weather, this was crazy talk. But nobody could find any mathematical or theoretical evidence to disprove it.

Much of Nima's work relates to the theory of the multiverse to which Maldacena's work on holograms alludes. In this model, as described by Nima's colleague Lisa Randall, our universe is just one of a nearly infinite number of universes floating through the soup of space-time like the bubbles in a glass of champagne. Each universe is a completely self-contained habitat—no particles or forces can go in or out—with one big exception: Gravity can travel freely, slipping from one membrane of strings, or universe, to another.

This is a big deal, because it offers a possible solution to what is called the hierarchical problem—why gravity is far, far weaker than the current theories might predict. (When compared with the electromagnetic and strong and weak nuclear forces, gravity appears to be ten million billion times weaker than it should be.) Nima's answer is as simple as it is astounding: The gravity that affects us has been deflected and diffused by other universes, like a bottle of whiskey that has been passed around the galaxy a few too many times. By the time the bottle reaches us, all that's left is a few fingers of backwash.

But here's the exciting part: Because Nima's proposed extra dimensions are so large, we may actually be able to see this cosmic shell game between universes in real time with the LHC—gravity from one dimension disappearing into the next. While a long shot by all accounts—including Nima's—this would be near irrefutable experimental evidence of string theory. The world, in fact, would be on a string.

And yet there's more. Because the most controversial ideas Nima hopes to test with the LHC don't deal with gravity but with the question of why there are 10^{500} possible string-theory solutions rather than just one. Unlike Maldacena and Witten, who believe that the near-infinite number of string theories may one day be reconciled into a single solution, Nima thinks each string solution could randomly apply to a different universe, and he hopes to prove it. It just happens that humans live in a universe tuned precisely to support life as we recognize it.

The idea is that each possible string theory, when coupled with a cosmological constant that varies randomly, corresponds to a different universe in the multiverse. The reason humans and all life exist in our universe is chance—the conditions just happened to be finely tuned in a way that allowed it. Termed the anthropic principle, it's a theory that drives many physicists insane, both because humans recapture their role at the center of our galaxy from Copernicus and because it seems utterly untestable: How could we ever test or even perceive the conditions in other universes if we're stuck in our own?

Nima may be in the minority, but he is undeterred. In fact, he's convinced that just as Maldacena showed how quantum mechanics could be used to show what happens in the formation and decay of black holes, quantum mechanics could help describe the contours of the multiverse we can't otherwise see. And just like many of the other aspects of string theory, the answer could stem from the LHC's experiments. It's a possible outcome that Witten acknowledges but despairs over. It would mean that science has finally jumped a barrier from being fully experimental to mostly theoretical.

Yet Nima is steadfast. "The mantra of string theory ten years ago was that the theory was smarter than you," he says. "So people would work on it, and there would always be more things coming out than went in. Well, exactly that—just follow the theory where it leads you, and it leads to this precipice. And now we have to decide what to do. So now a number of people are deciding to jump.... And I think that those of us that decided to take the plunge are staring at the true nature of the beast for the first time.

"I think this is the correct answer, and we are going to have to come to terms with it. And coming to terms with it is going to require a revolution of comparable magnitude to the revolution going from classical mechanics to quantum mechanics. So I think something similar is at stake. And the struggles we are having right now feel a lot like the sort of birthing pains in the twenty-five years from 1900 to 1925, when quantum mechanics started off as a twinkle in the eye of Planck and ended up as a full-fledged theory. I think we are sort of in the 1908, 1909 part of that period right now."

Which is to say, he thinks this moment is comparable to the most potent and revolutionary period in the history of physics, when Max Planck, Niels Bohr, and Einstein entirely changed the way we look at the atoms that make up our universe and our place in it.

SO THAT IS WHY Nima can't sleep. Much will be at stake when the LHC powers up. Supersymmetry, with those dreamed-about shadow particles string theorists have bet their careers on, could be a no-show, accelerating the end of string theory and giving renewed life to different ideas, like loop quantum gravity. Perimeter's unorthodox approach to research would be celebrated as prescient, Smolin and other vociferous critics of string theory remembered as visionaries with the guts to shout that the emperor has no clothes. Or else, of course, the pendulum and the Higgs boson could swing the way string theorists predict—and hope against all galactic hope—that it must. With a single selectron or graviton, M-theory and the holographic principle could begin their slow shift from the category of conjecture to principle; the theory of everything would be tantalizingly closer than ever before, our lonely place in the cosmos, trapped in an obscure two-dimensional universe within a sea of other two-dimensional universes, glimpsed for the first time.

Like Nima, we are all overlooking a steep precipice.

SYLVIA NASAR AND DAVID GRUBER

Manifold Destiny

FROM *THE NEW YORKER*

The Fields Medal, mathematics' Nobel Prize, was recently given to a Russian mathematician named Grigory Perelman for his resolution of a century-old problem. Perelman, however, refused to accept the honor. Sylvia Nasar and David Gruber report on this ascetic mathematician, and the international controversy that his proof provoked.

O N T H E E V E N I N G of June 20th, several hundred physicists, including a Nobel laureate, assembled in an auditorium at the Friendship Hotel in Beijing for a lecture by the Chinese mathematician Shing-Tung Yau. In the late nineteen-seventies, when Yau was in his twenties, he had made a series of breakthroughs that helped launch the string-theory revolution in physics and earned him, in addition to a Fields Medal—the most coveted award in mathematics—a reputation in both disciplines as a thinker of unrivalled technical power.

Yau had since become a professor of mathematics at Harvard and the director of mathematics institutes in Beijing and Hong Kong, dividing his time between the United States and China. His lecture at the Friendship Hotel was part of an international conference on string theory, which he had organized with the support of the Chinese government, in part to promote the country's recent advances in theoretical physics. (More than six thousand students attended the keynote address, which was delivered by Yau's close friend Stephen Hawking, in the Great Hall of the People.) The subject of Yau's talk was something that few in his audience knew much about: the Poincaré conjecture, a century-old conundrum about the characteristics of three-dimensional spheres, which, because it has important implications for mathematics and cosmology and because it has eluded all attempts at solution, is regarded by mathematicians as a holy grail.

Yau, a stocky man of fifty-seven, stood at a lectern in shirtsleeves and black-rimmed glasses and, with his hands in his pockets, described how two of his students, Xi-Ping Zhu and Huai-Dong Cao, had completed a proof of the Poincaré conjecture a few weeks earlier. "I'm very positive about Zhu and Cao's work," Yau said. "Chinese mathematicians should have every reason to be proud of such a big success in completely solving the puzzle." He said that Zhu and Cao were indebted to his longtime American collaborator Richard Hamilton, who deserved most of the credit for solving the Poincaré. He also mentioned Grigory Perelman, a Russian mathematician who, he acknowledged, had made an important contribution. Nevertheless, Yau said, "in Perelman's work, spectacular as it is, many key ideas of the proofs are sketched or outlined, and complete details are often missing." He added, "We would like to get Perelman to make comments. But Perelman resides in St. Petersburg and refuses to communicate with other people."

For ninety minutes, Yau discussed some of the technical details of his students' proof. When he was finished, no one asked any questions. That night, however, a Brazilian physicist posted a report of the lecture on his blog. "Looks like China soon will take the lead also in mathematics," he wrote.

GRIGORY PERELMAN IS INDEED RECLUSIVE. He left his job as a researcher at the Steklov Institute of Mathematics, in St. Petersburg, last December; he has few friends; and he lives with his mother in an apartment on the outskirts of the city. Although he had never granted an interview before, he was cordial and frank when we visited him, in late June, shortly after Yau's conference in Beijing, taking us on a long walking tour of the city. "I'm looking for some friends, and they don't have to be mathematicians," he said. The week before the conference, Perelman had spent hours discussing the Poincaré conjecture with Sir John M. Ball, the fifty-eight-year-old president of the International Mathematical Union, the discipline's influential professional association. The meeting, which took place at a conference center in a stately mansion overlooking the Neva River, was highly unusual. At the end of May, a committee of nine prominent mathematicians had voted to award Perelman a Fields Medal for his work on the Poincaré, and Ball had gone to St. Petersburg to persuade him to accept the prize in a public ceremony at the I.M.U.'s quadrennial congress, in Madrid, on August 22nd.

The Fields Medal, like the Nobel Prize, grew, in part, out of a desire to elevate science above national animosities. German mathematicians were excluded from the first I.M.U. congress, in 1924, and, though the ban was lifted before the next one, the trauma it caused led, in 1936, to the establishment of the Fields, a prize intended to be "as purely international and impersonal as possible."

However, the Fields Medal, which is awarded every four years, to between two and four mathematicians, is supposed not only to reward past achievements but also to stimulate future research; for this reason, it is given only to mathematicians aged forty and younger. In recent decades, as the number of professional mathematicians has grown, the Fields Medal has become increasingly prestigious. Only forty-four medals have been awarded in nearly seventy years—including three for work closely related to the Poincaré conjecture—and no mathematician has ever refused the prize. Nevertheless, Perelman told Ball that he had no intention of accepting it. "I refuse," he said simply.

Over a period of eight months, beginning in November, 2002, Perelman posted a proof of the Poincaré on the Internet in three installments. Like a sonnet or an aria, a mathematical proof has a distinct form and set of conventions. It begins with axioms, or accepted truths, and employs a series of logical statements to arrive at a conclusion. If the logic is deemed to be watertight, then the result is a theorem. Unlike proof in law or science, which is based on evidence and therefore subject to qualification and revision, a proof of a theorem is definitive. Judgments about the accuracy of a proof are mediated by peer-reviewed journals; to insure fairness, reviewers are supposed to be carefully chosen by journal editors, and the identity of a scholar whose paper is under consideration is kept secret. Publication implies that a proof is complete, correct, and original.

By these standards, Perelman's proof was unorthodox. It was astonishingly brief for such an ambitious piece of work; logic sequences that could have been elaborated over many pages were often severely compressed. Moreover, the proof made no direct mention of the Poincaré and included many elegant results that were irrelevant to the central argument. But, four years later, at least two teams of experts had vetted the proof and had found no significant gaps or errors in it. A consensus was emerging in the math community: Perelman had solved the Poincaré. Even so, the proof's complexity—and Perelman's use of shorthand in making some of his most important claims—made it vulnerable to challenge. Few mathematicians had the expertise necessary to evaluate and defend it.

After giving a series of lectures on the proof in the United States in 2003, Perelman returned to St. Petersburg. Since then, although he had continued to answer queries about it by e-mail, he had had minimal contact with colleagues and, for reasons no one understood, had not tried to publish it. Still, there was little doubt that Perelman, who turned forty on June 13th, deserved a Fields Medal. As Ball planned the I.M.U.'s 2006 congress, he began to conceive of it as a historic event. More than three thousand mathematicians would be attending, and King Juan Carlos of Spain had agreed to preside over the awards ceremony. The I.M.U.'s newsletter predicted that the congress would be remembered as "the occasion when this conjecture became a theorem." Ball, determined to make sure that Perelman would be there, decided to go to St. Petersburg.

Ball wanted to keep his visit a secret—the names of Fields Medal recipients are announced officially at the awards ceremony—and the conference center where he met with Perelman was deserted. For ten hours over two days, he tried to persuade Perelman to agree to accept the prize. Perelman, a slender, balding man with a curly beard, bushy eyebrows, and blue-green eyes, listened politely. He had not spoken English for three years, but he fluently parried Ball's entreaties, at one point taking Ball on a long walk—one of Perelman's favorite activities. As he summed up the conversation two weeks later: "He proposed to me three alternatives: accept and come; accept and don't come, and we will send you the medal later; third, I don't accept the prize. From the very beginning, I told him I have chosen the third one." The Fields Medal held no interest for him, Perelman explained. "It was completely irrelevant for me," he said. "Everybody understood that if the proof is correct then no other recognition is needed."

PROOFS OF THE POINCARÉ have been announced nearly every year since the conjecture was formulated, by Henri Poincaré, more than a hundred years ago. Poincaré was a cousin of Raymond Poincaré, the President of France during the First World War, and one of the most creative mathematicians of the nineteenth century. Slight, myopic, and notoriously absent-minded, he conceived his famous problem in 1904, eight years before he died, and tucked it as an offhand question into the end of a sixty-five-page paper.

Poincaré didn't make much progress on proving the conjecture.

"*Cette question nous entraînerait trop loin*" ("This question would take us too far"), he wrote. He was a founder of topology, also known as "rubber-sheet geometry," for its focus on the intrinsic properties of spaces. From a topologist's perspective, there is no difference between a bagel and a coffee cup with a handle. Each has a single hole and can be manipulated to resemble the other without being torn or cut. Poincaré used the term "manifold" to describe such an abstract topological space. The simplest possible two-dimensional manifold is the surface of a soccer ball, which, to a topologist, is a sphere—even when it is stomped on, stretched, or crumpled. The proof that an object is a so-called two-sphere, since it can take on any number of shapes, is that it is "simply connected," meaning that no holes puncture it. Unlike a soccer ball, a bagel is not a true sphere. If you tie a slipknot around a soccer ball, you can easily pull the slipknot closed by sliding it along the surface of the ball. But if you tie a slipknot around a bagel through the hole in its middle you cannot pull the slipknot closed without tearing the bagel.

Two-dimensional manifolds were well understood by the mid-nineteenth century. But it remained unclear whether what was true for two dimensions was also true for three. Poincaré proposed that all closed, simply connected, three-dimensional manifolds—those which lack holes and are of finite extent—were spheres. The conjecture was potentially important for scientists studying the largest known three-dimensional manifold: the universe. Proving it mathematically, however, was far from easy. Most attempts were merely embarrassing, but some led to important mathematical discoveries, including proofs of Dehn's Lemma, the Sphere Theorem, and the Loop Theorem, which are now fundamental concepts in topology.

By the nineteen-sixties, topology had become one of the most productive areas of mathematics, and young topologists were launching regular attacks on the Poincaré. To the astonishment of most mathematicians, it turned out that manifolds of the fourth, fifth, and higher dimensions were more tractable than those of the third dimension. By 1982, Poincaré's conjecture had been proved in all dimensions except the third. In 2000, the Clay Mathematics Institute, a private foundation that promotes mathematical research, named the Poincaré one of the seven most important outstanding problems in mathematics and offered a million dollars to anyone who could prove it.

"My whole life as a mathematician has been dominated by the Poincaré conjecture," John Morgan, the head of the mathematics department at Columbia University, said. "I never thought I'd see a solution. I thought nobody could touch it."

GRIGORY PERELMAN DID NOT PLAN to become a mathematician. "There was never a decision point," he said when we met. We were outside the apartment building where he lives, in Kupchino, a neighborhood of drab high-rises. Perelman's father, who was an electrical engineer, encouraged his interest in math. "He gave me logical and other math problems to think about," Perelman said. "He got a lot of books for me to read. He taught me how to play chess. He was proud of me." Among the books his father gave him was a copy of *Physics for Entertainment*, which had been a best-seller in the Soviet Union in the nineteen-thirties. In the foreword, the book's author describes the contents as "conundrums, brain-teasers, entertaining anecdotes, and unexpected comparisons," adding, "I have quoted extensively from Jules Verne, H. G. Wells, Mark Twain and other writers, because, besides providing entertainment, the fantastic experiments these writers describe may well serve as instructive illustrations at physics classes." The book's topics included how to jump from a moving car, and why, "according to the law of buoyancy, we would never drown in the Dead Sea."

The notion that Russian society considered worthwhile what Perelman did for pleasure came as a surprise. By the time he was fourteen, he was the star performer of a local math club. In 1982, the year that Shing-Tung Yau won a Fields Medal, Perelman earned a perfect score and the gold medal at the International Mathematical Olympiad, in Budapest. He was friendly with his teammates but not close—"I had no close friends," he said. He was one of two or three Jews in his grade, and he had a passion for opera, which also set him apart from his peers. His mother, a math teacher at a technical college, played the violin and began taking him to the opera when he was six. By the time Perelman was fifteen, he was spending his pocket money on records. He was thrilled to own a recording of a famous 1946 performance of *La Traviata*, featuring Licia Albanese as Violetta. "Her voice was very good," he said.

At Leningrad University, which Perelman entered in 1982, at the age of sixteen, he took advanced classes in geometry and solved a problem posed by Yuri Burago, a mathematician at the Steklov Institute, who later became his Ph.D. adviser. "There are a lot of students of high ability who speak before thinking," Burago said. "Grisha was different. He thought deeply. His answers were always correct. He always checked very, very carefully." Burago added, "He was not fast. Speed means nothing. Math doesn't depend on speed. It is about *deep*."

At the Steklov in the early nineties, Perelman became an expert on the geometry of Riemannian and Alexandrov spaces—extensions of traditional Euclidean geometry—and began to publish articles in the leading Russian and American mathematics journals. In 1992, Perelman was invited to spend a semester each at New York University and Stony Brook University. By the time he left for the United States, that fall, the Russian economy had collapsed. Dan Stroock, a mathematician at M.I.T., recalls smuggling wads of dollars into the country to deliver to a retired mathematician at the Steklov, who, like many of his colleagues, had become destitute.

Perelman was pleased to be in the United States, the capital of the international mathematics community. He wore the same brown corduroy jacket every day and told friends at N.Y.U. that he lived on a diet of bread, cheese, and milk. He liked to walk to Brooklyn, where he had relatives and could buy traditional Russian brown bread. Some of his colleagues were taken aback by his fingernails, which were several inches long. "If they grow, why wouldn't I let them grow?" he would say when someone asked why he didn't cut them. Once a week, he and a young Chinese mathematician named Gang Tian drove to Princeton, to attend a seminar at the Institute for Advanced Study.

For several decades, the institute and nearby Princeton University had been centers of topological research. In the late seventies, William Thurston, a Princeton mathematician who liked to test out his ideas using scissors and construction paper, proposed a taxonomy for classifying manifolds of three dimensions. He argued that, while the manifolds could be made to take on many different shapes, they nonetheless had a "preferred" geometry, just as a piece of silk draped over a dressmaker's mannequin takes on the mannequin's form.

Thurston proposed that every three-dimensional manifold could be broken down into one or more of eight types of component, including a spherical type. Thurston's theory—which became known as the geometrization conjecture—describes all possible three-dimensional manifolds and is thus a powerful generalization of the Poincaré. If it was confirmed, then Poincaré's conjecture would be, too. Proving Thurston and Poincaré "definitely swings open doors," Barry Mazur, a mathematician at Harvard, said. The implications of the conjectures for other disciplines may not be apparent for years, but for mathematicians the problems are fundamental. "This is a kind of twentieth-century Pythagorean theorem," Mazur added. "It changes the landscape."

In 1982, Thurston won a Fields Medal for his contributions to topology. That year, Richard Hamilton, a mathematician at Cornell, published a paper on an equation called the Ricci flow, which he suspected could be relevant for solving Thurston's conjecture and thus the Poincaré. Like a heat equation, which describes how heat distributes itself evenly through a substance—flowing from hotter to cooler parts of a metal sheet, for example—to create a more uniform temperature, the Ricci flow, by smoothing out irregularities, gives manifolds a more uniform geometry.

Hamilton, the son of a Cincinnati doctor, defied the math profession's nerdy stereotype. Brash and irreverent, he rode horses, wind-surfed, and had a succession of girlfriends. He treated math as merely one of life's pleasures. At forty-nine, he was considered a brilliant lecturer, but he had published relatively little beyond a series of seminal articles on the Ricci flow, and he had few graduate students. Perelman had read Hamilton's papers and went to hear him give a talk at the Institute for Advanced Study. Afterward, Perelman shyly spoke to him.

"I really wanted to ask him something," Perelman recalled. "He was smiling, and he was quite patient. He actually told me a couple of things that he published a few years later. He did not hesitate to tell me. Hamilton's openness and generosity—it really attracted me. I can't say that most mathematicians act like that.

"I was working on different things, though occasionally I would think about the Ricci flow," Perelman added. "You didn't have to be a great mathematician to see that this would be useful for geometrization. I felt I didn't know very much. I kept asking questions."

SHING-TUNG YAU WAS ALSO ASKING Hamilton questions about the Ricci flow. Yau and Hamilton had met in the seventies, and had become close, despite considerable differences in temperament and background. A mathematician at the University of California at San Diego who knows both men called them "the mathematical loves of each other's lives."

Yau's family moved to Hong Kong from mainland China in 1949, when he was five months old, along with hundreds of thousands of other refugees fleeing Mao's armies. The previous year, his father, a relief worker for the United Nations, had lost most of the family's savings in a series of failed ventures. In Hong Kong, to support his wife and eight children, he tutored college students in classical Chinese literature and philosophy.

When Yau was fourteen, his father died of kidney cancer, leaving his mother dependent on handouts from Christian missionaries and whatever small sums she earned from selling handicrafts. Until then, Yau had been an indifferent student. But he began to devote himself to schoolwork, tutoring other students in math to make money. "Part of the thing that drives Yau is that he sees his own life as being his father's revenge," said Dan Stroock, the M.I.T. mathematician, who has known Yau for twenty years. "Yau's father was like the Talmudist whose children are starving."

Yau studied math at the Chinese University of Hong Kong, where he attracted the attention of Shiing-Shen Chern, the preëminent Chinese mathematician, who helped him win a scholarship to the University of California at Berkeley. Chern was the author of a famous theorem combining topology and geometry. He spent most of his career in the United States, at Berkeley. He made frequent visits to Hong Kong, Taiwan, and, later, China, where he was a revered symbol of Chinese intellectual achievement, to promote the study of math and science.

In 1969, Yau started graduate school at Berkeley, enrolling in seven graduate courses each term and auditing several others. He sent half of his scholarship money back to his mother in China and impressed his professors with his tenacity. He was obliged to share credit for his first major result when he learned that two other mathematicians were working on the same problem. In 1976, he proved a twenty-year-old conjecture pertaining to a type of manifold that is now crucial to string theory. A French mathematician had formulated a proof of the problem, which is known as Calabi's conjecture, but Yau's, because it was more general, was more powerful. (Physicists now refer to Calabi-Yau manifolds.) "He was not so much thinking up some original way of looking at a subject but solving extremely hard technical problems that at the time only he could solve, by sheer intellect and force of will," Phillip Griffiths, a geometer and a former director of the Institute for Advanced Study, said.

In 1980, when Yau was thirty, he became one of the youngest mathematicians ever to be appointed to the permanent faculty of the Institute for Advanced Study, and he began to attract talented students. He won a Fields Medal two years later, the first Chinese ever to do so. By this time, Chern was seventy years old and on the verge of retirement. According to a relative of Chern's, "Yau decided that he was going to be the next famous Chinese mathematician and that it was time for Chern to step down."

Harvard had been trying to recruit Yau, and when, in 1983, it was about to make him a second offer Phillip Griffiths told the dean of faculty a version of a story from *The Romance of the Three Kingdoms*, a Chinese classic. In the third century A.D., a Chinese warlord dreamed of creating an empire, but the most brilliant general in China was working for a rival. Three times, the warlord went to his enemy's kingdom to seek out the general. Impressed, the general agreed to join him, and together they succeeded in founding a dynasty. Taking the hint, the dean flew to Philadelphia, where Yau lived at the time, to make him an offer. Even so, Yau turned down the job. Finally, in 1987, he agreed to go to Harvard.

Yau's entrepreneurial drive extended to collaborations with colleagues and students, and, in addition to conducting his own research, he began organizing seminars. He frequently allied himself with brilliantly inventive mathematicians, including Richard Schoen and William Meeks. But Yau was especially impressed by Hamilton, as much for his swagger as for his imagination. "I can have fun with Hamilton," Yau told us during the string-theory conference in Beijing. "I can go swimming with him. I go out with him and his girlfriends and all that." Yau was convinced that Hamilton could use the Ricci-flow equation to solve the Poincaré and Thurston conjectures, and he urged him to focus on the problems. "Meeting Yau changed his mathematical life," a friend of both mathematicians said of Hamilton. "This was the first time he had been on to something extremely big. Talking to Yau gave him courage and direction."

Yau believed that if he could help solve the Poincaré it would be a victory not just for him but also for China. In the mid-nineties, Yau and several other Chinese scholars began meeting with President Jiang Zemin to discuss how to rebuild the country's scientific institutions, which had been largely destroyed during the Cultural Revolution. Chinese universities were in dire condition. According to Steve Smale, who won a Fields for proving the Poincaré in higher dimensions, and who, after retiring from Berkeley, taught in Hong Kong, Peking University had "halls filled with the smell of urine, one common room, one office for all the assistant professors," and paid its faculty wretchedly low salaries. Yau persuaded a Hong Kong real-estate mogul to help finance a mathematics institute at the Chinese Academy of Sciences, in Beijing, and to endow a Fields-style medal for Chinese mathematicians under the age of forty-five. On his trips to China, Yau touted Hamilton and their joint work on the Ricci flow and the Poincaré as a model for young Chinese mathematicians. As he put it in Beijing, "They always say that the whole country should learn from Mao or some big heroes. So I made a joke to them, but I was half serious. I said the whole country should learn from Hamilton."

GRIGORY PERELMAN WAS LEARNING from Hamilton already. In 1993, he began a two-year fellowship at Berkeley. While he was there, Hamilton gave several talks on campus, and in one he mentioned that he was working on the Poincaré. Hamilton's Ricci-flow strategy was extremely technical and tricky to execute. After one of his talks at Berkeley, he told Perelman about his biggest obstacle. As a space is smoothed under the Ricci flow, some regions deform into what mathematicians refer to as "singularities." Some regions, called "necks," become attenuated areas of infinite density. More troubling to Hamilton was a kind of singularity he called the "cigar." If cigars formed, Hamilton worried, it might be impossible to achieve uniform geometry. Perelman realized that a paper he had written on Alexandrov spaces might help Hamilton prove Thurston's conjecture—and the Poincaré—once Hamilton solved the cigar problem. "At some point, I asked Hamilton if he knew a certain collapsing result that I had proved but not published—which turned out to be very useful,"

Perelman said. "Later, I realized that he didn't understand what I was talking about." Dan Stroock, of M.I.T., said, "Perelman may have learned stuff from Yau and Hamilton, but, at the time, they were not learning from him."

By the end of his first year at Berkeley, Perelman had written several strikingly original papers. He was asked to give a lecture at the 1994 I.M.U. congress, in Zurich, and invited to apply for jobs at Stanford, Princeton, the Institute for Advanced Study, and the University of Tel Aviv. Like Yau, Perelman was a formidable problem solver. Instead of spending years constructing an intricate theoretical framework, or defining new areas of research, he focussed on obtaining particular results. According to Mikhail Gromov, a renowned Russian geometer who has collaborated with Perelman, he had been trying to overcome a technical difficulty relating to Alexandrov spaces and had apparently been stumped. "He couldn't do it," Gromov said. "It was hopeless."

Perelman told us that he liked to work on several problems at once. At Berkeley, however, he found himself returning again and again to Hamilton's Ricci-flow equation and the problem that Hamilton thought he could solve with it. Some of Perelman's friends noticed that he was becoming more and more ascetic. Visitors from St. Petersburg who stayed in his apartment were struck by how sparsely furnished it was. Others worried that he seemed to want to reduce life to a set of rigid axioms. When a member of a hiring committee at Stanford asked him for a C.V. to include with requests for letters of recommendation, Perelman balked. "If they know my work, they don't need my C.V.," he said. "If they need my C.V., they don't know my work."

Ultimately, he received several job offers. But he declined them all, and in the summer of 1995 returned to St. Petersburg, to his old job at the Steklov Institute, where he was paid less than a hundred dollars a month. (He told a friend that he had saved enough money in the United States to live on for the rest of his life.) His father had moved to Israel two years earlier, and his younger sister was planning to join him there after she finished college. His mother, however, had decided to remain in St. Petersburg, and Perelman moved in with her. "I realize that in Russia I work better," he told colleagues at the Steklov.

At twenty-nine, Perelman was firmly established as a mathematician and yet largely unburdened by professional responsibilities. He was free to pursue whatever problems he wanted to, and he knew that his work, should he choose to publish it, would be shown serious consideration. Yakov Eliashberg, a mathematician at Stanford who knew Perelman at Berkeley, thinks that Perelman returned to Russia in order to work on the Poincaré. "Why not?" Perelman said when we asked whether Eliashberg's hunch was correct.

The Internet made it possible for Perelman to work alone while continuing to tap a common pool of knowledge. Perelman searched Hamilton's papers for clues to his thinking and gave several seminars on his work. "He didn't need any help," Gromov said. "He likes to be alone. He reminds me of Newton—this obsession with an idea, working by yourself, the disregard for other people's opinion. Newton was more obnoxious. Perelman is nicer, but very obsessed."

In 1995, Hamilton published a paper in which he discussed a few of his ideas for completing a proof of the Poincaré. Reading the paper, Perelman realized that Hamilton had made no progress on overcoming his obstacles—the necks and the cigars. "I hadn't seen any evidence of progress after early 1992," Perelman told us. "Maybe he got stuck even earlier." However, Perelman thought he saw a way around the impasse. In 1996, he wrote Hamilton a long letter outlining his notion, in the hope of collaborating. "He did not answer," Perelman said. "So I decided to work alone."

YAU HAD NO IDEA that Hamilton's work on the Poincaré had stalled. He was increasingly anxious about his own standing in the mathematics profession, particularly in China, where, he worried, a younger scholar could try to supplant him as Chern's heir. More than a decade had passed since Yau had proved his last major result, though he continued to publish prolifically. "Yau wants to be the king of geometry," Michael Anderson, a geometer at Stony Brook, said. "He believes that everything should issue from him, that he should have oversight. He doesn't like people encroaching on his territory." Determined to retain control over his field, Yau pushed his students to tackle big problems. At Harvard, he ran a notoriously tough seminar on differential geometry, which met for three hours at a time three times a week. Each student was assigned a recently published proof and asked to reconstruct it, fixing any errors and filling in gaps. Yau believed that a mathematician has an obligation to be explicit, and impressed on his students the importance of step-by-step rigor.

There are two ways to get credit for an original contribution in mathematics. The first is to produce an original proof. The second is to identify a significant gap in someone else's proof and supply the missing chunk. However, only true mathematical gaps—missing or mistaken arguments—can be the basis for a claim of originality. Filling in gaps in exposition—shortcuts and abbreviations used to make a proof more efficient—does not count. When, in 1993, Andrew Wiles revealed that a gap had been found in his proof of Fermat's last theorem, the problem became fair game for anyone, until, the following year, Wiles fixed the error. Most mathematicians would agree that, by contrast, if a proof's implicit steps can be made explicit by an expert, then the gap is merely one of exposition, and the proof should be considered complete and correct.

Occasionally, the difference between a mathematical gap and a gap in exposition can be hard to discern. On at least one occasion, Yau and his students have seemed to confuse the two, making claims of originality that other mathematicians believe are unwarranted. In 1996, a young geometer at Berkeley named Alexander Givental had proved a mathematical conjecture about mirror symmetry, a concept that is fundamental to string theory. Though other mathematicians found Givental's proof hard to follow, they were optimistic that he had solved the problem. As one geometer put it, "Nobody at the time said it was incomplete and incorrect."

In the fall of 1997, Kefeng Liu, a former student of Yau's who taught at Stanford, gave a talk at Harvard on mirror symmetry. According to two geometers in the audience, Liu proceeded to present a proof strikingly similar to Givental's, describing it as a paper that he had co-authored with Yau and another student of Yau's. "Liu mentioned Givental but only as one of a long list of people who had contributed to the field," one of the geometers said. (Liu maintains that his proof was significantly different from Givental's.)

Around the same time, Givental received an e-mail signed by Yau and his collaborators, explaining that they had found his arguments impossible to follow and his notation baffling, and had come up with a proof of their own. They praised Givental for his "brilliant idea" and wrote, "In the final version of our paper your important contribution will be acknowledged."

A few weeks later, the paper, "Mirror Principle I," appeared in the *Asian Journal of Mathematics*, which is co-edited by Yau. In it, Yau and his coauthors describe their result as "the first complete proof" of the mirror conjecture. They mention Givental's work only in passing. "Unfortunately," they write, his proof, "which has been read by many prominent experts, is incomplete." However, they did not identify a specific mathematical gap.

Givental was taken aback. "I wanted to know what their objection was," he told us. "Not to expose them or defend myself." In March, 1998, he published a paper that included a three-page footnote in which he pointed out a number of similarities between Yau's proof and his own. Several months later, a young mathematician at the University of Chicago who was asked by senior colleagues to investigate the dispute concluded that Givental's proof was complete. Yau says that he had been working on the proof for years with his students and that they achieved their result independently of Givental. "We had our own ideas, and we wrote them up," he says.

Around this time, Yau had his first serious conflict with Chern and the Chinese mathematical establishment. For years, Chern had been hoping to bring the I.M.U.'s congress to Beijing. According to several mathematicians who were active in the I.M.U. at the time, Yau made an eleventh-hour effort to have the congress take place in Hong Kong instead. But he failed to persuade a sufficient number of colleagues to go along with his proposal, and the I.M.U. ultimately decided to hold the 2002 congress in Beijing. (Yau denies that he tried to bring the congress to Hong Kong.) Among the delegates the I.M.U. appointed to a group that would be choosing speakers for the congress was Yau's most successful student, Gang Tian, who had been at N.Y.U. with Perelman and was now a professor at M.I.T. The host committee in Beijing also asked Tian to give a plenary address.

Yau was caught by surprise. In March, 2000, he had published a survey of recent research in his field studded with glowing references to Tian and to their joint projects. He retaliated by organizing his first conference on string theory, which opened in Beijing a few days before the math congress began, in late August, 2002. He persuaded Stephen Hawking and several Nobel laureates to attend, and for days the Chinese newspapers were full of pictures of famous scientists. Yau even managed to arrange for his group to have an audience with Jiang Zemin. A mathematician who helped organize the math congress recalls that along the highway between Beijing and the airport there were "billboards with pictures of Stephen Hawking plastered everywhere."

That summer, Yau wasn't thinking much about the Poincaré. He had confidence in Hamilton, despite his slow pace. "Hamilton is a very good friend," Yau told us in Beijing. "He is more than a friend. He is a hero. He is so original. We were working to finish our proof. Hamilton worked on it for twenty-five years. You work, you get tired. He probably got a little tired—and you want to take a rest."

Then, on November 12, 2002, Yau received an e-mail message from a Russian mathematician whose name didn't immediately register. "May I bring to your attention my paper," the e-mail said.

ON NOVEMBER 11TH, Perelman had posted a thirty-nine-page paper entitled "The Entropy Formula for the Ricci Flow and Its Geometric Applications," on arXiv.org, a Web site used by mathematicians to post preprints—articles awaiting publication in refereed journals. He then e-mailed an abstract of his paper to a dozen mathematicians in the United States—including Hamilton, Tian, and Yau—none of whom had heard from him for years. In the abstract, he explained that he had written "a sketch of an eclectic proof" of the geometrization conjecture.

Perelman had not mentioned the proof or shown it to anyone. "I didn't have any friends with whom I could discuss this," he said in St. Petersburg. "I didn't want to discuss my work with someone I didn't trust." Andrew Wiles had also kept the fact that he was working on Fermat's last theorem a secret, but he had had a colleague vet the proof before making it public. Perelman, by casually posting a proof on the Internet of one of the most famous problems in mathematics, was not just flouting academic convention but taking a considerable risk. If the proof was flawed, he would be publicly humiliated, and there would be no way to prevent another mathematician from fixing any errors and claiming victory. But Perelman said he was not particularly concerned. "My reasoning was: if I made an error and someone used my work to construct a correct proof I would be pleased," he said. "I never set out to be the sole solver of the Poincaré."

Gang Tian was in his office at M.I.T. when he received Perelman's e-mail. He and Perelman had been friendly in 1992, when they were both at N.Y.U. and had attended the same weekly math seminar in Princeton. "I immediately realized its importance," Tian said of Perelman's paper. Tian began to read the paper and discuss it with colleagues, who were equally enthusiastic.

On November 19th, Vitali Kapovitch, a geometer, sent Perelman an e-mail:

> Hi Grisha, Sorry to bother you but a lot of people are asking me about your preprint "The entropy formula for the Ricci ... " Do I understand it correctly that while you cannot yet do all the steps in the Hamilton program you can do enough so that using some collapsing results you can prove geometrization?

Perelman's response, the next day, was terse: "That's correct. Grisha."

In fact, what Perelman had posted on the Internet was only the first installment of his proof. But it was sufficient for mathematicians to see that he had figured out how to solve the Poincaré. Barry Mazur, the Harvard mathematician, uses the image of a dented fender to describe Perelman's achievement: "Suppose your car has a dented fender and you call a mechanic to ask how to smooth it out. The mechanic would have a hard time telling you what to do over the phone. You would have to bring the car into the garage for him to examine. Then he could tell you where to give it a few knocks. What Hamilton introduced and Perelman completed is a procedure that is independent of the particularities of the blemish. If you apply the Ricci flow to a 3-D space, it will begin to undent it and smooth it out. The mechanic would not need to even see the car—just apply the equation." Perelman proved that the "cigars" that had troubled Hamilton could not actually occur, and he showed that the "neck" problem could be solved by performing an intricate sequence of mathematical surgeries: cutting out singularities and patching up the raw edges. "Now we have a procedure to smooth things and, at crucial points, control the breaks," Mazur said.

Tian wrote to Perelman, asking him to lecture on his paper at M.I.T. Colleagues at Princeton and Stony Brook extended similar invitations. Perelman accepted them all and was booked for a month of lectures beginning in April, 2003. "Why not?" he told us with a shrug. Speaking of mathematicians generally, Fedor Nazarov, a mathematician at Michigan State University, said, "After you've solved a problem, you have a great urge to talk about it."

HAMILTON AND YAU WERE STUNNED by Perelman's announcement. "We felt that nobody else would be able to discover the solution," Yau told us in Beijing. "But then, in 2002, Perelman said that he published something. He basically did a shortcut without doing all the detailed estimates that we did." Moreover, Yau complained, Perel-

man's proof "was written in such a messy way that we didn't understand."

Perelman's April lecture tour was treated by mathematicians and
by the press as a major event. Among the audience at his talk at
Princeton were John Ball, Andrew Wiles, John Forbes Nash, Jr., who
had proved the Riemannian embedding theorem, and John Conway,
the inventor of the cellular automaton game Life. To the astonishment of many in the audience, Perelman said nothing about the Poincaré. "Here is a guy who proved a world-famous theorem and didn't
even mention it," Frank Quinn, a mathematician at Virginia Tech,
said. "He stated some key points and special properties, and then
answered questions. He was establishing credibility. If he had beaten
his chest and said, 'I solved it,' he would have got a huge amount of
resistance." He added, "People were expecting a strange sight. Perelman was much more normal than they expected."

To Perelman's disappointment, Hamilton did not attend that lecture or the next ones, at Stony Brook. "I'm a disciple of Hamilton's,
though I haven't received his authorization," Perelman told us. But
John Morgan, at Columbia, where Hamilton now taught, was in the
audience at Stony Brook, and after a lecture he invited Perelman to
speak at Columbia. Perelman, hoping to see Hamilton, agreed. The
lecture took place on a Saturday morning. Hamilton showed up late
and asked no questions during either the long discussion session that
followed the talk or the lunch after that. "I had the impression he had
read only the first part of my paper," Perelman said.

In the April 18, 2003, issue of *Science*, Yau was featured in an article
about Perelman's proof: "Many experts, although not all, seem convinced that Perelman has stubbed out the cigars and tamed the narrow
necks. But they are less confident that he can control the number of
surgeries. That could prove a fatal flaw, Yau warns, noting that many
other attempted proofs of the Poincaré conjecture have stumbled over
similar missing steps." Proofs should be treated with skepticism until
mathematicians have had a chance to review them thoroughly, Yau
told us. Until then, he said, "it's not math—it's religion."

By mid-July, Perelman had posted the final two installments of his proof on the Internet, and mathematicians had begun the work of formal explication, painstakingly retracing his steps. In the United States, at least two teams of experts had assigned themselves this task: Gang Tian (Yau's rival) and John Morgan; and a pair of researchers at the University of Michigan. Both projects were supported by the Clay Institute, which planned to publish Tian and Morgan's work as a book. The book, in addition to providing other mathematicians with a guide to Perelman's logic, would allow him to be considered for the Clay Institute's million-dollar prize for solving the Poincaré. (To be eligible, a proof must be published in a peer-reviewed venue and withstand two years of scrutiny by the mathematical community.)

On September 10, 2004, more than a year after Perelman returned to St. Petersburg, he received a long e-mail from Tian, who said that he had just attended a two-week workshop at Princeton devoted to Perelman's proof. "I think that we have understood the whole paper," Tian wrote. "It is all right."

Perelman did not write back. As he explained to us, "I didn't worry too much myself. This was a famous problem. Some people needed time to get accustomed to the fact that this is no longer a conjecture. I personally decided for myself that it was right for me to stay away from verification and not to participate in all these meetings. It is important for me that I don't influence this process."

In July of that year, the National Science Foundation had given nearly a million dollars in grants to Yau, Hamilton, and several students of Yau's to study and apply Perelman's "breakthrough." An entire branch of mathematics had grown up around efforts to solve the Poincaré, and now that branch appeared at risk of becoming obsolete. Michael Freedman, who won a Fields for proving the Poincaré conjecture for the fourth dimension, told the *Times* that Perelman's proof was a "small sorrow for this particular branch of topology." Yuri Burago said, "It kills the field. After this is done, many mathematicians will move to other branches of mathematics."

FIVE MONTHS LATER, Chern died, and Yau's efforts to insure that he—not Tian—was recognized as his successor turned vicious. "It's all about their primacy in China and their leadership among the expatriate Chinese," Joseph Kohn, a former chairman of the Princeton mathematics department, said. "Yau's not jealous of Tian's mathematics, but he's jealous of his power back in China."

Though Yau had not spent more than a few months at a time on mainland China since he was an infant, he was convinced that his status as the only Chinese Fields Medal winner should make him Chern's successor. In a speech he gave at Zhejiang University, in Hangzhou, during the summer of 2004, Yau reminded his listeners of his Chinese roots. "When I stepped out from the airplane, I touched the soil of Beijing and felt great joy to be in my mother country," he said. "I am proud to say that when I was awarded the Fields Medal in mathematics, I held no passport of any country and should certainly be considered Chinese."

The following summer, Yau returned to China and, in a series of interviews with Chinese reporters, attacked Tian and the mathematicians at Peking University. In an article published in a Beijing science newspaper, which ran under the headline "SHING-TUNG YAU IS SLAMMING ACADEMIC CORRUPTION IN CHINA," Yau called Tian "a complete mess." He accused him of holding multiple professorships and of collecting a hundred and twenty-five thousand dollars for a few months' work at a Chinese university, while students were living on a hundred dollars a month. He also charged Tian with shoddy scholarship and plagiarism, and with intimidating his graduate students into letting him add his name to their papers. "Since I promoted him all the way to his academic fame today, I should also take responsibility for his improper behavior," Yau was quoted as saying to a reporter, explaining why he felt obliged to speak out.

In another interview, Yau described how the Fields committee had passed Tian over in 1988 and how he had lobbied on Tian's behalf with various prize committees, including one at the National Science

Foundation, which awarded Tian five hundred thousand dollars in 1994.

Tian was appalled by Yau's attacks, but he felt that, as Yau's former student, there was little he could do about them. "His accusations were baseless," Tian told us. But, he added, "I have deep roots in Chinese culture. A teacher is a teacher. There is respect. It is very hard for me to think of anything to do."

While Yau was in China, he visited Xi-Ping Zhu, a protégé of his who was now chairman of the mathematics department at Sun Yat-sen University. In the spring of 2003, after Perelman completed his lecture tour in the United States, Yau had recruited Zhu and another student, Huai-Dong Cao, a professor at Lehigh University, to undertake an explication of Perelman's proof. Zhu and Cao had studied the Ricci flow under Yau, who considered Zhu, in particular, to be a mathematician of exceptional promise. "We have to figure out whether Perelman's paper holds together," Yau told them. Yau arranged for Zhu to spend the 2005–06 academic year at Harvard, where he gave a seminar on Perelman's proof and continued to work on his paper with Cao.

ON APRIL 13TH of this year, the thirty-one mathematicians on the editorial board of the *Asian Journal of Mathematics* received a brief e-mail from Yau and the journal's co-editor informing them that they had three days to comment on a paper by Xi-Ping Zhu and Huai-Dong Cao titled "The Hamilton-Perelman Theory of Ricci Flow: The Poincaré and Geometrization Conjectures," which Yau planned to publish in the journal. The e-mail did not include a copy of the paper, reports from referees, or an abstract. At least one board member asked to see the paper but was told that it was not available. On April 16th, Cao received a message from Yau telling him that the paper had been accepted by the *A.J.M.*, and an abstract was posted on the journal's Web site.

A month later, Yau had lunch in Cambridge with Jim Carlson, the president of the Clay Institute. He told Carlson that he wanted to trade a copy of Zhu and Cao's paper for a copy of Tian and Morgan's book manuscript. Yau told us he was worried that Tian would try to steal from Zhu and Cao's work, and he wanted to give each party simultaneous access to what the other had written. "I had a lunch with Carlson to request to exchange both manuscripts to make sure that nobody can copy the other," Yau said. Carlson demurred, explaining that the Clay Institute had not yet received Tian and Morgan's complete manuscript.

By the end of the following week, the title of Zhu and Cao's paper on the *A.J.M.*'s Web site had changed, to "A Complete Proof of the Poincaré and Geometrization Conjectures: Application of the Hamilton-Perelman Theory of the Ricci Flow." The abstract had also been revised. A new sentence explained, "This proof should be considered as the crowning achievement of the Hamilton-Perelman theory of Ricci flow."

Zhu and Cao's paper was more than three hundred pages long and filled the *A.J.M.*'s entire June issue. The bulk of the paper is devoted to reconstructing many of Hamilton's Ricci-flow results—including results that Perelman had made use of in his proof—and much of Perelman's proof of the Poincaré. In their introduction, Zhu and Cao credit Perelman with having "brought in fresh new ideas to figure out important steps to overcome the main obstacles that remained in the program of Hamilton." However, they write, they were obliged to "substitute several key arguments of Perelman by new approaches based on our study, because we were unable to comprehend these original arguments of Perelman which are essential to the completion of the geometrization program." Mathematicians familiar with Perelman's proof disputed the idea that Zhu and Cao had contributed significant new approaches to the Poincaré. "Perelman already did it and what he did was complete and correct," John Morgan said. "I don't see that they did anything different."

By early June, Yau had begun to promote the proof publicly. On

June 3rd, at his mathematics institute in Beijing, he held a press conference. The acting director of the mathematics institute, attempting to explain the relative contributions of the different mathematicians who had worked on the Poincaré, said, "Hamilton contributed over fifty per cent; the Russian, Perelman, about twenty-five per cent; and the Chinese, Yau, Zhu, and Cao et al., about thirty per cent." (Evidently, simple addition can sometimes trip up even a mathematician.) Yau added, "Given the significance of the Poincaré, that Chinese mathematicians played a thirty-per-cent role is by no means easy. It is a very important contribution."

On June 12th, the week before Yau's conference on string theory opened in Beijing, the *South China Morning Post* reported, "Mainland mathematicians who helped crack a 'millennium math problem' will present the methodology and findings to physicist Stephen Hawking. . . . Yau Shing-Tung, who organized Professor Hawking's visit and is also Professor Cao's teacher, said yesterday he would present the findings to Professor Hawking because he believed the knowledge would help his research into the formation of black holes."

On the morning of his lecture in Beijing, Yau told us, "We want our contribution understood. And this is also a strategy to encourage Zhu, who is in China and who has done really spectacular work. I mean, important work with a century-long problem, which will probably have another few century-long implications. If you can attach your name in any way, it is a contribution."

E. T. Bell, the author of *Men of Mathematics*, a witty history of the discipline published in 1937, once lamented "the squabbles over priority which disfigure scientific history." But in the days before e-mail, blogs, and Web sites, a certain decorum usually prevailed. In 1881, Poincaré, who was then at the University of Caen, had an altercation with a German mathematician in Leipzig named Felix Klein. Poincaré had published several papers in which he labelled certain functions "Fuchsian," after another mathematician. Klein wrote to Poincaré, pointing out that he and others had done significant work on these functions, too. An exchange of polite letters between Leipzig and Caen ensued. Poincaré's last word on the subject was a quote from Goethe's "Faust": "*Name ist Schall und Rauch.*" Loosely translated, that corresponds to Shakespeare's "What's in a name?"

This, essentially, is what Yau's friends are asking themselves. "I find myself getting annoyed with Yau that he seems to feel the need for more kudos," Dan Stroock, of M.I.T., said. "This is a guy who did magnificent things, for which he was magnificently rewarded. He won every prize to be won. I find it a little mean of him to seem to be trying to get a share of this as well." Stroock pointed out that, twenty-five years ago, Yau was in a situation very similar to the one Perelman is in today. His most famous result, on Calabi-Yau manifolds, was hugely important for theoretical physics. "Calabi outlined a program," Stroock said. "In a real sense, Yau was Calabi's Perelman. Now he's on the other side. He's had no compunction at all in taking the lion's share of credit for Calabi-Yau. And now he seems to be resenting Perelman getting credit for completing Hamilton's program. I don't know if the analogy has ever occurred to him."

Mathematics, more than many other fields, depends on collaboration. Most problems require the insights of several mathematicians in order to be solved, and the profession has evolved a standard for crediting individual contributions that is as stringent as the rules governing math itself. As Perelman put it, "If everyone is honest, it is natural to share ideas." Many mathematicians view Yau's conduct over the Poincaré as a violation of this basic ethic, and worry about the damage it has caused the profession. "Politics, power, and control have no legitimate role in our community, and they threaten the integrity of our field," Phillip Griffiths said.

Perelman likes to attend opera performances at the Mariinsky Theatre, in St. Petersburg. Sitting high up in the back of the house, he can't make out the singers' expressions or see the details of their costumes. But he cares only about the sound of their voices, and he says that the acoustics are better where he sits than anywhere else in the theatre. Perelman views the mathematics community—and much of the larger world—from a similar remove.

Before we arrived in St. Petersburg, on June 23rd, we had sent several messages to his e-mail address at the Steklov Institute, hoping to arrange a meeting, but he had not replied. We took a taxi to his apartment building and, reluctant to intrude on his privacy, left a book—a collection of John Nash's papers—in his mailbox, along with a card saying that we would be sitting on a bench in a nearby playground the following afternoon. The next day, after Perelman failed to appear, we left a box of pearl tea and a note describing some of the questions we hoped to discuss with him. We repeated this ritual a third time. Finally, believing that Perelman was out of town, we pressed the buzzer for his apartment, hoping at least to speak with his mother. A woman answered and let us inside. Perelman met us in the dimly lit hallway of the apartment. It turned out that he had not checked his Steklov e-mail address for months, and had not looked in his mailbox all week. He had no idea who we were.

We arranged to meet at ten the following morning on Nevsky Prospekt. From there, Perelman, dressed in a sports coat and loafers, took us on a four-hour walking tour of the city, commenting on every building and vista. After that, we all went to a vocal competition at the St. Petersburg Conservatory, which lasted for five hours. Perelman repeatedly said that he had retired from the mathematics community and no longer considered himself a professional mathematician. He mentioned a dispute that he had had years earlier with a collaborator over how to credit the author of a particular proof, and said that he was dismayed by the discipline's lax ethics. "It is not people who break ethical standards who are regarded as aliens," he said. "It is people like me who are isolated." We asked him whether he had read Cao and Zhu's paper. "It is not clear to me what new contribution did they make," he said. "Apparently, Zhu did not quite understand the argument and reworked it." As for Yau, Perelman said, "I can't say I'm outraged. Other people do worse. Of course, there are many mathematicians who are more or less honest. But almost all of them are conformists. They are more or less honest, but they tolerate those who are not honest."

The prospect of being awarded a Fields Medal had forced him to make a complete break with his profession. "As long as I was not conspicuous, I had a choice," Perelman explained. "Either to make some ugly thing"—a fuss about the math community's lack of integrity—"or, if I didn't do this kind of thing, to be treated as a pet. Now, when I become a very conspicuous person, I cannot stay a pet and say nothing. That is why I had to quit." We asked Perelman whether, by refusing the Fields and withdrawing from his profession, he was eliminating any possibility of influencing the discipline. "I am not a politician!" he replied, angrily. Perelman would not say whether his objection to awards extended to the Clay Institute's million-dollar prize. "I'm not going to decide whether to accept the prize until it is offered," he said.

Mikhail Gromov, the Russian geometer, said that he understood Perelman's logic: "To do great work, you have to have a pure mind. You can think only about the mathematics. Everything else is human weakness. Accepting prizes is showing weakness." Others might view Perelman's refusal to accept a Fields as arrogant, Gromov said, but his principles are admirable. "The ideal scientist does science and cares about nothing else," he said. "He wants to live this ideal. Now, I don't think he really lives on this ideal plane. But he wants to."

Robin Marantz Henig

Looking for the Lie

FROM THE *NEW YORK TIMES MAGAZINE*

There are many techniques to try to determine if someone is lying, from the polygraph to face-reading, and currently the U. S. government is looking for a more reliable machine to detect lies. Exploring the researchers studying lying, Robin Marantz Henig begins to suspect a foolproof lie detector might change human society, and not necessarily for the better.

WHEN PEOPLE HEAR that I'm writing an article about deception, they're quick to tell me how to catch a liar. Liars always look to the left, several friends say; liars always cover their mouths, says a man sitting next to me on a plane. Beliefs about how lying looks are plentiful and often contradictory: depending on whom you choose to believe, liars can be detected because they fidget a lot, hold very still, cross their legs, cross their arms, look up, look down, make eye contact or fail to make eye contact. Freud thought anyone could spot deception by paying close enough attention, since the liar, he wrote, "chatters with his fingertips; betrayal oozes out of him at every pore." Nietzsche wrote that "the mouth may lie, but the face it makes nonetheless tells the truth."

This idea is still with us, the notion that liars are easy to spot. Just last month, Charles Bond, a psychologist at Texas Christian University, reported that among 2,520 adults surveyed in 63 countries, more than 70 percent believe that liars tend to avert their gazes. The majority also believe that liars squirm, stutter, touch or scratch themselves or tell longer stories than usual. The liar stereotype exists in just about every culture, Bond wrote, and its persistence "would be less puzzling if we had more reason to imagine that it was true." What is true, instead, is that there are as many ways to lie as there are liars; there's no such thing as a dead giveaway.

Most people think they're good at spotting liars, but studies show otherwise. A very small minority of people, probably fewer than 5 percent, seem to have some innate ability to sniff out deception with accuracy. But in general, even professional lie-catchers, like judges and customs officials, perform, when tested, at a level not much better than chance. In other words, even the experts would have been right almost as often if they had just flipped a coin.

In the middle of the war on terrorism, the federal government is not willing to settle for 50-50 odds. "Credibility assessment" is the new catch phrase, which emerged at about the same time as "red-level alert" and "homeland security." Unfortunately, most of the devices now available, like the polygraph, detect not the lie but anxiety about the lie. The polygraph measures physiological responses to stress, like increases in blood pressure, respiration rate and electrodermal skin response. So it can miss the most dangerous liars: the ones who don't care that they're lying, don't know that they're lying or have been trained to lie. It can also miss liars with nothing to lose if they're detected, the true believers willing to die for the cause.

Responding to federal research incentives, a handful of scientists are building a cognitive theory of deception to show what lying looks like—on a liar's face, in a liar's demeanor and, most important, in a liar's brain. The ultimate goal is a foolproof technology for deception detection: a brain signature of lying, something as visible and unambiguous as Pinocchio's nose.

Deception is a complex thing, evanescent and difficult to pin down; it's no accident that the poets describe it with diaphanous imagery like "tangled web" and "tissue of lies." But the federal push for a new device for credibility assessment leaves little room for complexity; the government is looking for a blunt instrument, a way to pick out black and white from among the duplicitous grays.

Nearly a century ago the modern polygraph started out as a machine in search of an application; it hung around for lack of anything better. But the polygraph has been mired in controversy for years, with no strong scientific theory to adequately explain why, or even whether, it works. If the premature introduction of a new machine is to be avoided this time around, the first step is to do something that was never done with the polygraph, to develop a theory of the neurobiology of deception. Two strands of scientific work are currently involved in this effort: brain mapping, which uses the 21st century's most sophisticated techniques for visualizing patterns of brain metabolism and electrical activity; and face reading, which uses tools that are positively prehistoric, the same two eyes used by our primate ancestors to spot a liar.

As these two strands, the ancient and the futuristic, contribute to a new generation of lie detectors, the challenge will be twofold: to resist pressure to introduce new technologies before they are adequately tested and to fight the overzealous use of these technologies in places where they do not belong—to keep inviolable that most private preserve of our ordinary lives, the place inside everyone's head where secrets reside.

THE ENGLISH LANGUAGE HAS 112 words for deception, according to one count, each with a different shade of meaning: collusion, fakery, malingering, self-deception, confabulation, prevarication, exaggeration, denial. Lies can be verbal or nonverbal, kindhearted or self-serving, devious or baldfaced; they can be lies of omission or lies of commission; they can be lies that undermine national security or lies that make a child feel better. And each type might involve a unique neural pathway.

To develop a theory of deception requires parsing the subject into its most basic components so it can be studied one element at a time. That's what Daniel Langleben has been doing at the University of Pennsylvania. Langleben, a psychiatrist, started an experiment on deception in 2000 with a simple design: a spontaneous yes-no lie using a deck of playing cards. His research involved taking brain images with a functional-M.R.I. scanner, a contraption not much bigger than a kayak but weighing 10 tons. Unlike a traditional M.R.I., which provides a picture of the brain's anatomy, the functional M.R.I. shows the brain in action. It takes a reading, every two to three seconds, of how much oxygen is being used throughout the brain, and that information is superimposed on an anatomical brain map to determine which regions are most active while performing a particular task.

There's very little about being in a functional-M.R.I. scanner that is natural: you are flat on your back, absolutely still, with your head immobilized by pillows and straps. The scanner makes a dreadful din, which headphones barely muffle. If you're part of an experiment, you might be given a device with buttons to press for "yes" or "no" and another device with a single panic button. Not only is the physical setup unnatural, but in most deception studies the experimental design is unnatural, too. It is difficult to replicate the real-world conditions of lying—the relationship between liar and target, the urgency not to get caught—in a functional-M.R.I. lab, or in any other kind of lab. But as an early step in mapping the lying brain, such artificiality has to suffice.

In Langleben's first deception study at Penn, the subjects were told at the beginning of the experiment to lie about a particular playing card, the five of clubs. To be sure the card carried no emotional weight, Langleben screened out compulsive gamblers from the group. One at a time, the subjects lay motionless in the scanner, watched pictures of playing cards flash onto a screen and pressed a button indicating whether they had that card or not. When an image of a card they didn't have came up, the subjects, as they had been instructed, told the truth and pressed "no." But when an image of the five of clubs came up, they also pressed "no," even though the card was in their pockets. That is, whenever they saw the five of clubs, they lied.

According to Langleben, certain regions of the brain were more active on average when his 18 subjects were lying than when they were telling the truth. Lying was associated with increased activity in several areas of the cortex, including the anterior cingulate cortex and the superior frontal gyrus. "We didn't have a map of deception in the brain—we still don't—so we didn't know exactly what this meant," Langleben said. "But that wasn't the question we were asking at the time in any case. What we were asking with that first experiment was, 'Can the difference in brain activity between lie and truth be detected by functional M.R.I.?' Our study showed that it can." He said that the prefrontal cortex—the reasoning part of the brain—was generally more aroused during lying than during truth-telling, an indication that it took more cognitive work to lie.

Brain mappers are just beginning to figure out how different parts of the brain function. The function of one region found to be activated in the five-of-clubs experiment, the anterior cingulate cortex, is still the subject of some debate; it is thought, among other things, to help a person choose between two conflicting responses, which makes it a logical place to look for a signature of deception. This region is also activated during the Stroop task, in which a series of words are written in different colors and the subject must respond with what color the ink is, disregarding the word itself. This is harder than it sounds, at least when the written word is a color word that is different from the ink it is written in. If the word "red" is written in blue, for instance, a lot of people say "red" instead of "blue." Telling a spontaneous lie is similar to the Stroop task in that it involves holding two things in mind simultaneously—in this case, the truth and the lie—and making a choice about which one to apply.

Langleben performed his card experiment again in 2003, with a few refinements, including giving his subjects the choice of two cards to lie about and whether to lie at all. This second study found activation in some of the same regions as the first, establishing a pattern of deception-related activity in particular parts of the cortex: one in the front, two on the sides and two in the back. The finding in the back, the parietal cortex, intrigued Langleben.

"At first I thought the parietal finding was a fluke," he said. The parietal cortex is usually activated during arousal of various kinds. It is also involved in the manifestation of thoughts as physical changes, like goose bumps that erupt when you're afraid, or sweating that increases when you lie. The connection to sweating interested Langleben, since sweating is also one of the polygraph's hallmark measurements. He looked at existing studies of this response, and in all of them he found activity that could be traced back to the parietal lobe. Until Langleben's observation of its connection to brain changes, the sweat response (which the polygraph measures with sensors on the palm or fingertips) had been thought to be a purely "downstream" change, a secondary effect caused not by the lie itself but by the consequences of lying: guilt, anxiety, fear or the excess positive emotion one researcher calls "duping delight." But Langleben's findings indicated that it might have a corollary "upstream," in the central nervous system. This meant that at least one polygraph measurement might have a signature right at the source of the lie, the brain itself.

So there it was: the first intimation of a Pinocchio response.

The parietal-cortex finding, while speculative, is "interesting to pay attention to because of its relationship to the polygraph," Langleben said. "In this way, we might not have to cancel the polygraph. We may be able to put it on firm neuroscience footing."

———

OVER AT HARVARD, Stephen Kosslyn, a psychologist, was looking at the map Langleben was starting to build and found himself troubled by the connection between deception and the anterior cingulate cortex. "Yes, it lights up during spontaneous lying," Kosslyn said, but it also lights up during other tasks, like the Stroop task, that have nothing to do with deception. "So it couldn't be the lie zone." Deception "is a huge, multidimensional space," he said, "in which every combination of things matters." Kosslyn began by thinking about the different dimensions, the various ways that lies differ from one another in terms of how they are produced. Is the lie about you, or about someone else? Is it about something you did yesterday or something your friend plans to do tomorrow? Do you feel strongly about the lie? Are there serious consequences to getting caught? Each type of lie might lead to activation of particular parts of the brain, since each type involves its own set of neural processes.

He decided to compare the brain tracings for lies that are spontaneous, like those in Langleben's study, with those that are rehearsed. A spontaneous lie comes when a mother asks her teenage son, "Did you do your math homework?" A rehearsed lie comes when she asks him, "Why are you coming home an hour past your curfew?" The question about the homework probably surprises him, and he has to lie on the fly. The question about the curfew was probably one he had been anticipating, and concocting an answer to, for most of the previous hour.

Kosslyn's working hypothesis was that different brain networks are used during spontaneous lying than are used during truth-telling or the telling of a memorized lie. Spontaneous lying requires the liar not only to generate the lie and keep the lie in mind but also to keep in mind what the truth is, to avoid revealing it by mistake. In contrast, Kosslyn said, a rehearsed lie requires only that an individual retrieve the lie from memory, since the work of establishing a credible lie has already been done.

To help his subjects generate meaningful lies to memorize, Kosslyn asked them to provide details about one notable work experience and one vacation experience. Then he helped them construct what he called an "alternative-reality scenario" about one of them. (The other experience he held in reserve as the basis for his subject's unrehearsed spontaneous lies.) If the experience was a vacation in Miami, for instance, Kosslyn changed it to San Diego; if the person had gone there to visit a sister, he changed it to a visit to Uncle Sol. Kosslyn had the participants practice the false scenario for a few hours, and then he put them into a scanner at Harvard's functional-M.R.I. facility. There were 10 subjects altogether, all in their 20's.

As he predicted, Kosslyn found that as far as the brain was concerned, spontaneous and rehearsed lies were two different things. They both involved memory processing, but of different kinds of memories, which in turn activated different regions of the cortex: one part of the frontal lobe (involved in working memory) for the spontaneous lie; a different part in the right anterior frontal cortex (involved in retrieving episodic memory) for the lie that was rehearsed. That's not much of a map yet, but it is a cumulative movement toward a theory of deception: that lying involves different cognitive work than truth-telling and that it activates several regions in the cerebral cortex that are also activated during certain memory and thinking tasks.

Even as these small bits of data emerge through functional-M.R.I. imagery, however, Kosslyn remains skeptical about the brain-mapping enterprise as a whole. "If I'm right, and deception turns out to be not just one thing, we need to start pulling the bird apart by its joints and looking at the underlying systems involved," he said. A true understanding of deception requires a fuller knowledge of functions like memory, perception and visual imagery, he said, aspects of neuroscience investigations not directly related to deception at all.

In Kosslyn's view, brain mapping and lie detection are two different things. The first is an academic exercise that might reveal some basic information about how the brain works, not only during lying but also during other high-level tasks; it uses whatever technology is available in the sophisticated neurophysiology lab. The second is a real-world enterprise, best accomplished not necessarily by using elaborate instruments but by encouraging people "to use their two eyes and brains." Searching for a "lie zone" of the brain as a counter-terrorism strategy, he said, is like trying to get to the moon by climbing a tree. It feels as if you're getting somewhere because you're moving higher and higher. But then you get to the top of the tree, and there's nowhere else to go, and the moon is still hundreds of thousands of miles away. Better to have stayed on the ground and really figured out the problem before setting off on a path that looks like progress but is really nothing more than motion. Better, in this case, to discover what deception looks like in the brain by breaking it down into progressively smaller elements, no matter how artificial the setup and how tedious the process, before introducing a lie-detection device that doesn't really get you where you want to go.

EVEN THE MOST ENTHUSIASTIC BRAIN MAPPERS probably agree with one aspect of Kosslyn's skeptical analysis: a true brain map of lying is, at best, elusive. Part of the difficulty comes from the technology itself. In the world of brain mapping, a functional-M.R.I. scan paints a picture that is broad and, in its way, lumbering. It can indicate which region of the brain is active, but it can take a reading no more frequently than once every two seconds.

For a more refined picture of cognitive change from one instant to the next, scientists have turned to the electroencephalogram, which detects neural impulses on the scale of milliseconds. But while EEG's might be ideal to answer "when" questions about brain activity, they are not so good at answering questions about "where." Most EEG's use 10 or 12 electrodes attached by a tacky glue at scattered spots on the scalp, which record electrical impulses firing from the brain as a whole. They give little indication of which region is doing the firing.

That's why deception researchers use a refined version of the ordinary EEG, which increases the number of electrodes from 12 to 128. These 128 electrodes, each the size of a typewriter key, are studded around a stretchy mesh cap. Using the cap, investigators can trace where electrical impulses are coming from when a person lies.

The cap is unwieldy and uncomfortable—definitely not ready yet for the world outside the laboratory. Jennifer Vendemia, a psychologist at the University of South Carolina, has been using the cap since 2000, when she began studying deception by looking at a particular class of brain wave known as E.R.P., for event-related potential. The E.R.P. wave represents electrical activity in response to a stimulus, usually 300 or 400 milliseconds after the stimulus is shown. It can be a sign that high-level cognitive processes, like paying attention and retrieving memories, are taking place.

Vendemia has studied deception and E.R.P. waves in 626 undergraduates. She outfits them with the electrode cap and a plastic barbershop cape, which is necessary because, in order to maintain an electrical circuit, each of the 128 electrodes has to be thoroughly soaked. The cap is sopping wet when she puts it on her subjects, and during the experiment Vendemia occasionally comes into the room with a squirter and soaks it down some more.

Vendemia presented her subjects with a series of true-false statements, like "The grass is green" and "A snake has 13 legs," which they were instructed to answer either truthfully or deceptively, depending on which color the statement was written in. The subjects took a longer time—up to 200 milliseconds longer, on average—to lie than to tell the truth. They revealed a change in certain E.R.P. waves while they were lying, especially in the regions of the brain that the functional-M.R.I. scanners also focused on as possible lie zones: the parietal and medial regions of the brain, along the top and middle of the head.

"E.R.P. has the advantage of being a little more portable, and substantially less expensive, than M.R.I.," Vendemia said. "But E.R.P. cannot do some of the things that functional M.R.I. can do. If you're trying to model the brain, you really need both techniques."

One thing E.R.P. might eventually be able to do is predict whether someone intends to lie—even before he or she has made a decision about it. This brings us into sci-fi territory, into the realm of mind reading. When Vendemia has a subject in an E.R.P. cap, she can detect the first brain-wave changes within 240 to 260 milliseconds after a true-false statement appears on a computer screen. But these changes are an indication of intention, not action; it can take 400 to 600 milliseconds for a person to decide whether to respond with "true" or "false." "With E.R.P., I've taken away your right to make a decision about your response," Vendemia said. "It's the ultimate invasion." If someone knows before you do what your brain is indicating as your intention, is there any room left, in that window of a few hundred milliseconds, for the exercise of free will? Or have you already been labeled a liar by your spontaneous brain waves, without your having a chance to override them and choose a different path?

Lies make secrets possible; they let us carve out a private territory that no one, not even those closest to us, can enter without our permission. Without lies, there can be no such sanctuary, no interior life that is completely and inviolably ours. Do we want to allow anyone, whether a government interrogator or a beloved spouse, unfettered access to that interior life?

Even a practiced lie-catcher like Paul Ekman recognizes that lying is a matter of privacy. "I don't use my ability to spot lies in my personal life," said Ekman, emeritus professor of psychology at the University of California, San Francisco. If his wife or two grown children want to lie to him, he said, that's their business: "They haven't given me the right to call them on their lies."

In his book *Telling Lies*, Ekman underscored this point. His Facial Action Coding System, a precise categorization of the 10,000 or so expressions that are created by various combinations of 43 independent muscles in the face, allows him to do the same kind of mind reading that Vendemia can do with her E.R.P. cap. Facial expressions are hard-wired into the brain, according to Ekman, and can erupt without an individual's awareness about 200 milliseconds after a stimulus. Much like E.R.P. waves, then, a facial expression can give away your feelings before you are even aware of them, before you have made a conscious decision about whether to lie about those feelings or not. "Detecting clues to deceit is a presumption," Ekman wrote. "It takes without permission, despite the other person's wishes."

But in many situations, it's important to know who's lying to you, whether the liar wants you to or not. And for those times, Ekman said, his system of lie detection can be taught to anyone, with an accuracy rate of more than 95 percent. His holistic perspective is almost the polar opposite of brain mappers like Langleben's and Vendemia's: instead of focusing on the liar's neurons, Ekman takes a long, hard look at the liar's face.

The Facial Action Coding System is the key to Ekman's strategy. Basic emotions lead to characteristic facial expressions, which only a handful of really good liars manage to conceal. Part of lying is putting on a false face that's consistent with the lie. But even practiced liars, according to Ekman, may not always be able to control the "leakage" of their true feelings, which flit across the face in microexpressions that last less than half a second. These microexpressions indicate an incongruity between the liar's words and his emotions. "It doesn't mean he's lying necessarily," Ekman said. "It's what I call a 'hot spot,' a point of discontinuity that deserves investigation."

Ekman teaches police investigators, embassy officials and others how to spot liars, including how to read these microexpressions. He begins by showing photos of faces in apparently neutral poses. In each face, a microexpression appears for 40 milliseconds, and the trainee has to press a button to indicate which emotion was in that microexpression: fear, anger, surprise, happiness, sadness, contempt or disgust. When I took the pretest to measure my innate lie-detecting capabilities, I could see the microexpressions in about 70 percent of the examples. But after about 15 minutes of training, I improved. The training session let me stop the action if I missed a question, since Ekman's idea is that if you know what you're looking for—and the microexpressions, when frozen, are vivid and easy to name—you can spot them even when they flash by in an instant. In the post-training test, I scored an 86 percent.

In addition to microexpressions, Ekman said, certain aspects of a person's demeanor can indicate whether he is lying. Voice, hand movements, posture, speech patterns: when these vary from how the person usually speaks or gesticulates, or when they don't fit the situation, that's another hot spot to explore. Word choices often change with lying, too, with the speaker using "distancing language," like fewer first-person pronouns and more in the third person. Also common are what Ekman calls "verbal hedges," which liars might use to buy time as they figure out what they want to say. To illustrate a verbal hedge, Ekman pointed to one of the many cartoons he uses in his workshops: a shark standing in a courtroom, looking up at the judge and saying, "Define 'frenzy.' "

Ekman enjoys using these insights to unmask the lies of public figures (though he has a rule that prohibits him from commenting on any elected official currently in office, no matter how tempting a target). At his home in the Oakland Hills, he has a videotape library of some of the most notable lies of recent history, and he showed me how to watch one when I visited last fall. It was from a presidential news conference in early 1998, during the first days of the Monica Lewinsky scandal. Ekman smiled as he watched it; he knows this clip well. "I want you to listen to me," President Bill Clinton was saying, shaking his forefinger like a schoolmarm. "I did not have sexual relations with that woman."

There it was: the president's "distancing language," calling Lewinsky "that woman," and an almost imperceptible softening of his voice at the end of the sentence. When this news conference was originally broadcast, Ekman said, "everyone I had ever trained from all over the country called me and said: 'Did you see the president? He's lying.' "

Even though Ekman has been hired to teach his technique to embassy workers and military intelligence officers—to the tune of $35,000 for a five-day workshop—his low-tech approach to lie-catching is definitely out of vogue. "After 9/11," he said, "I contacted different federal agencies—the Defense Department, the C.I.A.—and said, 'I think there are some things I can teach your agents that can be of help right now.' " But several turned him down, he said, with one person bluntly stating, "I can't support anything unless it ends in a machine doing it."

THE QUEST FOR SUCH A MACHINE has roots in the early 20th century, when the first modern lie detector, a rudimentary polygraph, was introduced. The man often cited as its inventor, William Moulton Marston, was a Harvard-trained psychologist who went on to make his mark as the creator of the comic-book character Wonder Woman. Not coincidentally, one of Wonder Woman's most potent weapons was her Magic Lasso, which made it impossible for anyone in its grip to tell a lie.

Marston spent 20 years trying to get his machine used by the military, in courts and even in advertising. After the success of Wonder Woman, however, he used it mostly for entertainment. His comic-book editor, Sheldon Mayer, recalled being hooked up to a polygraph during a party at Marston's home. After a few warm-up questions, Marston tossed him a zinger, "Do you think you're the greatest cartoonist in the world?"

As Mayer wrote in his memoir, "I felt I was being quite truthful when I said no, and it turned out I was lying!" What an interesting reaction—even if, as was likely, Mayer was just trying to be funny. Because how prescient, really, to joke that the machine must have been right, that the machine knew more about Mayer than he did himself. It's the power of a simple mechanical device to make you doubt your own concept of truth and lie—"It turned out I was lying"—that made the polygraph so alluring, and so disturbing. And it's that power, combined with the idea that the machines are peering directly into the brain, that makes the polygraph's modern counterparts even more so.

Today, the polygraph is the subject of much controversy, with organizations devoted to publicizing "countermeasures"—ways to subvert the results—to prove how unreliable it is. But the American Polygraph Association says it has "great probative value," and police departments still use it to help focus their criminal investigations and to try to extract confessions. The polygraph is also used to screen potential and current federal employees in law enforcement and for security clearances, although private employers are prohibited from using it as a pre-employment screen. Polygraphists are also routinely brought in to investigate such matters as insurance fraud, corporate theft and contested divorce.

But there is little scientific evidence to back up the accuracy of the polygraph. "There has been no serious effort in the U.S. government to develop the scientific basis for the psychophysiological detection of deception by any technique," stated a report issued by the National Research Council in 2003. Polygraph research has been "managed and supported by national security and law enforcement agencies that do not operate in a culture of science," the council said, suggesting that these are not the best settings for an objective assessment of any device's pros and cons.

The polygraph has many cons. It requires a suspect who is cooperative, feels guilty or anxious about lying and hasn't been educated to the various countermeasures that can thwart the results. Polygraph results can be more reliable in investigations in which the questioners already know what they're looking for. This allows investigators to develop a line of questioning that leads to something like the Guilty Knowledge Test. This is a multiple-choice test in which the answer is something only a guilty person would know—and only a guilty person's polygraph readings would indicate arousal upon hearing it.

The history of polygraphs is a cautionary tale, an example of how not to introduce the next generation of credibility-assessment devices. "Security and law enforcement agencies need to improve their capability to independently evaluate claims proffered by advocates of new techniques for detecting deception," the National Research Council said. "The history of the polygraph makes clear that such agencies typically let clinical judgment outweigh scientific evidence."

History is in some danger of repeating itself at the site of the government's most focused effort to look for the next generation of lie detectors, the Department of Defense Polygraph Institute. This is where the brain mapping of the academic investigators is turned into practical machinery. Scientists at Dodpi (pronounced DOD-pie) are an inventive bunch, investigating instruments that measure the body's emission of heat, light, vibration or any other physiological properties that might change when someone tells a lie.

The Dodpi facility sits at one end of the huge Army base at Fort Jackson, S.C., where Army recruits en route to Iraq go for basic training. Among the new machines being studied is a thermal scanner, in which a computer image of a person's face is color-coded according to how much heat it emits. The region of interest, just inside each eye, grows hotter when a person lies. It also grows hotter during many other cognitive tasks, however, so a more specific signature for deception might be required to keep the thermal scanner from falling prey to the same problems of imprecision as the polygraph.

Another machine is the eye tracker, which follows a person's gaze—its fixation, duration, rapid eye movements and scanning path—to determine if he's looking at something he has seen before. It can be thought of as a mute version of the Guilty Knowledge Test.

Other high-tech deception detectors—many of them capable of remote operation, so they could theoretically be used without a suspect's knowledge—are being developed at laboratories across the country, with financing from agencies like Dodpi, the Department of Homeland Security and the Defense Advanced Research Projects Agency. (Defense Department officials will not reveal the amount they spend on credibility assessment, nor the degree to which the budget has increased since 9/11, because some of the research is classified.) The detectors look for increases in physiological processes that are associated with lying: a sniffer test that measures levels of stress hormones on the breath, for instance, a pupillometer that measures pupil dilation and a near-infrared-light beam that measures blood flow to the cerebral cortex.

With this push for an automated lie detector, some observers worry that we'll see a replay of the polygraph experience: the marketing of a halfway technology not quite capable of separating lying from other cognitive or emotional tasks. The polygraph was a machine in search of an application, and it became entrenched in criminal justice more out of habit than out of proved efficacy. This could easily happen again, as credibility assessment is being lauded as a crucial counterterrorism tool.

"The fear is that because so much money has been put into homeland security, that people may be trying to find quick solutions to complex problems by buying something," said Tom Zeffiro, a neurologist at Georgetown University and chairman of a workshop on high-tech credibility assessment sponsored last summer by the National Science Foundation. "And technology that might not be thoroughly evaluated might be put into practice." Already there are efforts to sell computer algorithms and devices that some scientists believe to be insufficiently tested, products with names like Brain Fingerprinting and No Lie M.R.I. Zeffiro said that one of his workshop suggestions is to establish a neutral testing laboratory to keep such products from being used commercially before there is at least some minimum amount of evidence that they work.

Big Brother concerns hover in the background, too, with some of these instruments, especially the smallest ones. It is sobering to think that we might be moving toward a society in which hidden sensors are trying, in one way or another, to read our minds. At Dodpi, however, scientists don't seem to fret much about such things. "The operational use of what we develop is not something we think about," said Andrew Ryan, a former police psychologist who is the head of research at Dodpi. "Our job is to develop the science. Once that science is developed, how it's used is up to other people."

EACH DAY WE WALK a fine line between deception and discretion. "Everybody lies," Mark Twain wrote, "every day; every hour; awake; asleep; in his dreams; in his joy; in his mourning."

First there are the lies of omission. You go out to dinner with your sister and her handsome new boyfriend, and you find him obnoxious.

When you and your sister discuss the evening later, isn't it a lie for you to talk about the restaurant and not about the boyfriend? What if you talk about his good looks and not about his offensive personality?

Then there are the lies of commission, many of which are harmless, the lies that allow us to get along with one another. When you receive a gift you can't use, or are invited to lunch with a co-worker you dislike, you're likely to say, "Thank you, it's perfect" or "I wish I could, but I have a dentist's appointment," rather than speak the harsher truth. These are the lies we teach our children to tell; we call them manners. Even our automatic response of "Fine" to a neighbor's equally automatic "How are you?" is often, when you get right down to it, a lie.

More serious lies can have a range of motives and implications. They can be malicious, like lying about a rival's behavior in order to get him fired, or merely strategic, like not telling your wife about your mistress. Not every one of them is a lie that needs to be uncovered. "We humans are active, creative mammals who can represent what exists as if it did not, and what doesn't exist as if it did," wrote David Nyberg, a visiting scholar at Bowdoin College, in *The Varnished Truth*. "Concealment, obliqueness, silence, outright lying—all help to hold Nemesis at bay; all help us abide too-large helpings of reality."

Learning to lie is an important part of maturation. What makes a child able to start telling lies, usually at about age 3 or 4, is that he has begun developing a theory of mind, the idea that what goes on in his head is different from what goes on in other people's heads. With his first lie to his mother, the power balance shifts imperceptibly: he now knows something she doesn't know. With each new lie, he gains a bit more power over the person who believes him. After a while, the ability to lie becomes just another part of his emotional landscape.

"Lying is just so ordinary, so much a part of our everyday lives and everyday conversations," that we hardly notice it, said Bella DePaulo, a psychologist at the University of California, Santa Barbara. "And in many cases it would be more difficult, challenging and stressful for people to tell the truth than to lie."

In the 1990's, DePaulo asked 147 people to keep a diary of their social interactions for one week and to note "any time you intentionally try to mislead someone," either verbally or nonverbally. At the end of the week, the subjects had lied, on average, 1.5 times a day. "Lied about where I had been," read a diary entry. "Said that I did not have change for a dollar." "Told him I had done poorly on my calculus homework when I had aced it." "Said I had been true to my girl."

People didn't feel guilty about these lies, by and large, but lying still left them with what DePaulo called a "smudge," a sort of smarmy feeling after lying. Her subjects reported feeling less positive about their interactions with people to whom they had lied than to people to whom they had not lied.

Still, DePaulo said that her research led her to believe that not all lying is bad, that it often serves a perfectly respectable purpose; in fact, it is sometimes a nobler, or at least kinder, option than telling the truth. "I call them kindhearted lies, the lies you tell to protect someone else's life or feelings," DePaulo said. A kindhearted lie is when a genetic counselor says nothing when she happens to find out, during a straightforward test for birth defects, that a man could not possibly have fathered his wife's new baby. It's when a neighbor lies about hiding a Jewish family in Nazi-occupied Poland. It's when a doctor tells a terminally ill patient that the new chemotherapy might work. And it's when a mother tells her daughter that nothing bad will ever happen to her.

"We found in our studies that these were the lies that women most often told to other women," DePaulo said. "Women are the ones saying, 'You did the right thing,' 'I know just how you feel,' 'That was a lovely dinner,' 'You look great.' I don't think they're doing that because they think the truth is unimportant or because they have a casual attitude toward lying. I think they just value their friends' feelings more than they value the truth."

If the search for an all-purpose lie detector were successful, and these everyday lies were uncovered along with the threatening or malicious ones, we might, paradoxically, end up feeling a little less safe than we felt before. It would be destabilizing indeed to be stripped of the half-truths and delusions on which social life depends.

PERSONALLY, I cannot tell a lie. This is not to say that I never lie; I'm just not very good at it. My lies are mostly lies of omission, secrets that I choose not to talk about at all, because when I do say something deceptive, people usually see right through me.

I realize that my honesty comes as much from ineptitude as from integrity. In fact, my own dirty little secret is that I wish I were a better liar; I think it would make me a more interesting person, maybe even a better writer. Still, when I told Paul Ekman that I hardly ever lie, I detected in my voice an unseemly amount of pride.

Ekman's response surprised me. He is, after all, one of the nation's leading experts on spotting liars; I expected him to nod sagely, approving of me and my truthful ways. Instead, he listened to my veiled boast with a patient little smile. Then, without missing a beat, he started enumerating the qualities that are required to lie. To lie, he told me, a person needs three things: to be able to think strategically and plan her moves ahead of time, like a good chess player; to read the needs of other people and put herself in their shoes, like a good therapist; and to manage her emotions, like a grown-up person.

So had this very nice, polite, accomplished man dissed me? Had Ekman told me, indirectly, that bad liars like me are immature, unempathetic and not especially bright? Had he pointed out that the skills of lying are the same skills involved in the best human social interactions?

Probably. (Would he tell me the truth if I asked him?) Deception is, after all, one trait associated with the evolution of higher intelligence. According to the Machiavellian Intelligence Hypothesis, developed by Richard Byrne and Andrew Whiten, two Scottish primatologists at the University of St. Andrews in Fife, the more social a species, the more intelligent it is. The hypothesis holds that as social interactions became more and more complex, our primate ancestors evolved so they could engage in the trickery, manipulation, skulduggery and sleight of hand needed to live in large social groups, which helped them avoid predators and survive.

"All of a sudden, the idea that intelligence began in social manipulation, deceit and cunning cooperation seems to explain everything we had always puzzled about," Byrne and Whiten wrote. In 2004, Byrne and another colleague, Nadia Corp, looked at the brains and behavior of 18 primate species and found empirical support for the hypothesis: the bigger the neocortex, the more deceptive the behavior.

But even if liars are smarter and more successful than the rest of us, most people I know seem to be like me: secretly smug about their honesty, rarely admitting to telling lies. One notable exception is a friend who blurted out, when she heard I was writing this article, "I lie all the time."

Lying started early for this friend, she wrote later in an e-mail message. She grew up in a big, competitive family (14 siblings and stepsiblings in all), and lying was the easiest way to get a word in edgewise. "There were so many of us," she wrote, "asserting knowledge became even more prized than the knowledge itself—because you were heard." If she didn't know something, she made it up. It got her siblings' attention.

These days, my friend, who is a novelist, claims that she lies almost reflexively. Maybe it gives her power to have information she's not sharing with her loved ones, she wrote; maybe it's just something left over from her childhood struggle for attention and from her writerly need to flex her imagination. "If I'm on the phone and my husband walks in and says, 'Who's that?' I might say Jill if it's Joan," she wrote to me. "Or if my mother asks me who I had lunch with, I'll tell her Caroline when it was Alice, for no good reason. This kind of useless lying is harmless except when I get caught—and I do get caught, because keeping track of useless lies is both daunting and exhausting."

My friend might be aware of getting caught, but habitual liars like her are probably harder to spot than neophytes like me. People who lie a lot are good at it, and they don't worry as much as I do about getting caught. And no matter what device or technique of lie detection is used, the liar who doesn't strain at her deception is still less likely to be fingered than the liar who does.

In a recent study looking at the brain anatomy of pathological liars versus nonliars, researchers at the University of Southern California found that the liars had more white matter in their prefrontal cortexes. The investigators found their subjects partly through self-identification, an odd choice in a study of pathological liars. But it was an intriguing finding nonetheless. "White matter is pivotal to the connectivity and cognitive function of the human brain," Sean Spence, a deception researcher and psychiatrist at the University of Sheffield, wrote in an editorial accompanying the study's publication in the *British Journal of Psychiatry* last October. "And abnormal prefrontal white matter might affect complex behaviors such as deception."

As for my lying friend, she may or may not have an excess of white matter in the front of her brain. But she does lie, often. Her lies have no hidden purpose, she told me; she lies only for the sake of lying. "Of course," she added in her e-mail message, "I could be lying about all of this."

Today's federal effort to develop an efficient machine for credibility assessment has been compared to the Manhattan Project, the secret government undertaking to build the atomic bomb. This sounds hyperbolic, to compare a high-tech lie detector to a weapon of mass destruction, but Tom Zeffiro, who made the analogy, said that they raise similar moral quandaries, especially for the scientists doing the research. If a truly efficient lie detector could be developed, he said, we might find ourselves living in "a fundamentally different world than the one we live in today."

In the quest to make the country safer by looking for brain tracings of lies, it might turn out to be all but impossible to tell which tracings are signatures of truly dangerous lies and which are the images of lies that are harmless and kindhearted, or self-serving without being dangerous. As a result, we might find ourselves with instruments that can detect deception not only as an antiterrorism device but also in situations that have little to do with national security: job interviews, tax audits, classrooms, boardrooms, courtrooms, bedrooms.

This would be a problem. As the great physician-essayist Lewis Thomas once wrote, a foolproof lie-detection device would turn our quotidian lives upside down: "Before long, we would stop speaking to each other, television would be abolished as a habitual felon, politicians would be confined by house arrest and civilization would come to a standstill." It would be a mistake to bring such a device too rapidly to market, before considering what might happen not only if it didn't work—which is the kind of risk we're accustomed to thinking about—but also what might happen if it did. Worse than living in a world plagued by uncertainty, in which we can never know for sure who is lying to whom, might be to live in a world plagued by its opposite: certainty about where the lies are, thus forcing us to tell one another nothing but the truth.

JOSHUA DAVIS

Face Blind

FROM *WIRED*

Some people are good with faces, some are bad with faces, and some, strange to say, cannot perceive faces at all. Living with a condition called prosopagnosia—face blindness—they cannot recognize people, or even themselves, by facial characteristics. Joshua Davis follows an Internet-based community of the face blind and the unlikely researcher who studies them.

BILL CHOISSER WAS 48 when he first recognized himself. He was standing in his bathroom, looking in the mirror when it happened. A strand of hair fell down—he had been growing it out for the first time. The strand draped toward a nose. He understood that it was a nose, but then it hit him forcefully that it was *his* nose. He looked a little higher, stared into his own eyes, and saw . . . himself.

For most of his childhood, Choisser thought he was normal. He just assumed that nobody saw faces. But slowly, it dawned on him that he was different. Other people recognized their mothers on the street. He did not. During the 1970s, as a small-town lawyer in the Illinois Ozarks, he struggled to convince clients that he was competent even though he couldn't find them in court. He never greeted the judges when he passed them on the street—everyone looked similarly blank to him—and he developed a reputation for arrogance. His father, also a lawyer, told him to pay more attention. His mother grew distant from him. He felt like he lived in a ghost world. Not being able to see his own face left him feeling hollow.

One day in 1979, he quit, left town, and set out to find a better way of being in the world. At 32, he headed west and landed a job as a number cruncher at a construction firm in San Francisco. The job isolated him—he spent his days staring at formulas—but that was a good thing: He didn't have to talk to people much. With 1,500 miles between him and southern Illinois, he felt a measure of freedom. He started to wear colorful bandannas, and he let his hair grow. When it got long enough, he found that it helped him see himself. Before that, he'd had to deduce his presence: I'm the only one in the room, so that must be me in the mirror. Now that he had long hair and a wild-looking scarf on his head, he could recognize his image. He felt the beginnings of an identity.

It gave him the confidence to start seeing doctors. He wanted to know if there was something wrong with his brain. His vision was fine, they told him—20/20. One doctor suggested he might have emotional problems and referred him to a psychiatrist. In the medical literature, there were a few reports of head-injury and stroke victims who'd lost their ability to recognize faces. No one, as far as the doctors knew, had ever been born with the condition.

Conventional medicine, in other words, got him nowhere. So Choisser posted a message about his experiences on a Usenet group devoted to people with neurological problems. His subject line was "Trouble Recognizing Faces." After a few months, in late 1996, he received a solitary reply. "Hello, Bill," the e-mail began, "I read what you wrote, and I think I have what you have."

BRAD DUCHAINE WAS SO LOADED, he didn't realize he was plummeting to the ground. It was his 27th birthday party and the handsome Wisconsin native had been trying to have a good time. He'd had more than a few beers, grappled onto the roof to enjoy the beautiful Santa Barbara, California, sky, and fallen asleep. Now, inexplicably, he was falling. He smashed into the concrete driveway and his hip shattered. Luckily, the beer dulled the pain. He clawed his way to the living room and fell asleep among a few other unconscious revelers. Though no one knew it at the time, this guy—passed out, busted up, bleeding on the floor—was going to have a lasting impact on neuroscience.

Duchaine had started out enjoying himself too much as an undergrad and nearly flunked out of two colleges. He tried again at Marquette University in Milwaukee, and this time pulled it together. In the cold, boring Midwest nights, he settled down and did well, graduating summa cum laude in 1995. That got him into the cognitive psychology PhD program at UC Santa Barbara.

But the fall off the roof two years into his PhD coincided with a crack in his confidence. His dissertation was in shambles. He hadn't been able to find a suitable topic and was playing videogames until his thumbs hurt. Did he really think he could succeed as a neuroscientist? His parents were both midlevel managers at a paper company back in Wisconsin. Community college instructor was a more realistic goal.

But when Duchaine's parents visited their son in the summer of 1998, he didn't tell them any of this. The couple had invited an old friend to dinner at the Shoreline Beach Café, a seaside restaurant with tables right on the sand, and it didn't seem like the right time to reveal his doubts. Instead, over appetizers, talk turned to a presentation on unusual neurological cases that Duchaine had given to undergraduates earlier in the day. He was especially fascinated by a study of people who'd been hit on the head and lost their sense of direction or ability to perceive certain objects.

"I know a kid who doesn't recognize faces," the friend said.

"He had a stroke?" Duchaine asked. He knew that in a few rare cases—usually as a result of stroke or trauma—patients reported that they had lost the ability to recognize faces. Oliver Sacks' "man who mistook his wife for a hat" was the most famous example.

"No, he's always been that way," the friend replied.

Duchaine called the teenager's family a few days later. The boy's father told him that if he really wanted to learn more about the condition, he should check out a Yahoo group and Web site maintained by a guy in San Francisco named Bill Choisser. A whole community of these people were chatting online—that's how the teen discovered he wasn't the only one who had a problem with faces. Choisser had even begun to popularize a name for the condition: face blindness. If this were true—if there was a large, previously unidentified population with the condition—it would be a major discovery. Duchaine started to think he might make it as a neuroscientist after all.

IN OCTOBER 1944, as the Russian Red Army advanced into Germany, a single artillery round struck a German command post on the East Prussian front. Everyone inside was ripped apart by shrapnel, but a 36-year-old lieutenant somehow lived. He was evacuated to a field hospital, where a surgeon removed a piece of metal from the back of his head. The external wound healed, but it soon became clear that something strange was going on. The officer reported that he could no longer see faces.

He was sent to a psychiatric hospital near Stuttgart, where a doctor named Joachim Bodamer examined him. There had been reports of face blindness as far back as antiquity, but no one had studied it systematically, so the physician decided to make a detailed analysis.

To assess the extent of the man's impairment, Bodamer dressed the officer's wife of seven years as a nurse and lined her up with four real nurses. Bodamer asked if he noticed anything different about any of the nurses. The man said no. Next, he was told to look in a mirror and report what he saw. "It's strange," he said. "I've looked at myself often, but that's not me anymore, although I know that it's me. But I have a feeling of unfamiliarity."

Bodamer wrote a 47-page report on the case and coined a name for the condition: prosopagnosia (in Greek, *prosopo* is face and *agnosia* means without knowledge). He defined it as "the selective disruption of the perception of faces, one's own face as well as those of others, which are seen but not recognized as faces belonging to a particular owner."

While he was producing the report, the world around him was going up in flames. The Allies dropped more than 7,000 tons of bombs on Dresden, incinerating the city. Stuttgart—13 miles from Bodamer's home—came next. As his work on the paper drew to a close, fundamental questions about human consciousness weighed on him. What might his patient's condition imply about how healthy people experience the world? Maybe there are mental mechanisms for perceiving each aspect of reality—parts of the human mind that render faces, houses, and bombs meaningful. While everything was falling apart, Bodamer had stumbled onto a clue about how things come together in our heads. Perhaps the fluid reality we perceive is just a flimsy construct of individual puzzle pieces, any one of which could suddenly disappear. In nothing else, Bodamer wrote, "does medical fact touch so closely on . . . the basis of all knowledge."

OVER THE NEXT FIVE DECADES, researchers trying to pursue these questions had one major problem: Not many people get shrapnel shot though their brain and survive. In other words, the pool of potential research subjects was minuscule. Scientists did discover that a small number of stroke victims developed prosopagnosia, but their lesions often continued to grow, preventing reliable comparisons and, in many cases, causing death.

Still, these patients tantalized researchers. "People think of the brain as one seamless intelligence, because that's how it feels to us from the inside," says Nancy Kanwisher, an MIT vision researcher. "But if you look at prosopagnosia, you start to realize that the brain may actually be a grouping of stand-alone computational machines that are wired together."

The question is, how many machines are there and how do they work? On a basic level, we know that there are distinct visual, auditory, memory, and motor systems. But within each of these, are there further specializations? For instance, within the visual cortex, is there a specialized part of the brain devoted to faces? And, if so, are there further specializations within that for gender, skin color, age, and even attractiveness? It's akin to discovering the molecule, only to realize that there may be atoms, electrons, and quarks as well.

As functional MRI came into use in the 1990s, neuroscientists rolled people under the magnets to track blood flow to specific parts of their brain. They learned that a small part of the visual cortex— eventually dubbed the fusiform face area—showed increased blood activity whenever a subject looked at a face. Researchers had discovered what appeared to be a computational machine devoted specifically to processing faces. Stroke or gunshot victims tended to develop prosopagnosia if they sustained injuries to that area. But that was about as much as this relatively crude imaging technique could reveal.

The best that scientists had been able to do was point to a particular part of the brain and say, "That's important." The face-processing system remained a black box.

WHEN ZOË HUNN WAS 14, her three closest friends decided to enter a modeling contest in a London department store. The girls tried to convince Hunn to sign up, too. She thought it was a silly idea; looks didn't matter to her, and she had no idea whether she was pretty. She had never paid much attention to her face—it didn't seem to represent who she was.

Though she didn't know it, Hunn was severely face blind. Her father had the same problem. Both just assumed that they were bad with faces, in the same way some people are bad with names. They developed elaborate coping strategies, like focusing on voices and searching for clues in a conversation. Inevitably, they embarrassed themselves.

Since all her friends were entering the contest, Hunn decided to go along. To her surprise, she ended up winning the top prize: an offer from a modeling agency. In 2002, she was signed by Models 1 in London, the same agency that represents Stephanie Seymour and Linda Evangelista. She appeared in *Vogue* and *Elle* and in fashion ads across Europe, earning as much as $1,500 a day. But she could never spot herself in any of the photos. She might recognize the clothes and deduce her presence in the picture, but she was never sure.

Hunn decided to see a doctor. She explained to him that she was a rapidly rising model—this should be the time of her life. It wasn't. She was completely unable to appreciate her beauty, which had now become the centerpiece of her young life. When she should have been going out to parties and having fun, she chose to stay home. "Everyone looks the same," she told the doctor, "so it's hard to connect emotionally with anyone." The doctor checked her eyes, made sure she didn't have a tumor, and then recommended counseling for shyness.

In the summer of 2003, she traveled to Edinburgh, Scotland, for the annual theater festival. On the third night, she saw a performer who was unusually memorable. He was a tall mime with white hair and vivid black eyebrows. She stared at him. He was the first person she felt she'd ever really seen.

Later that night, the unimaginable happened: Hunn recognized him in a bar. It was like being thrown a lifeline. She mustered the courage to introduce herself and told him that his performance made her laugh. He smiled and thanked her. She learned his name was Mick, and that was all she needed. She was in love. It didn't matter that he was a 38-year-old mime trying to make ends meet. She could *see* him.

Mick, for his part, was captivated not by her beauty but by the way she watched him as if her whole world depended on the sight of him. It was a performer's dream, and Mick melted in the intensity of it.

Despite the protests of her parents, they moved in together within a few months.

One day in their new home, she read a short article in a British magazine about a man named Brad Duchaine and his work on something called face blindness. It described her symptoms: inability to recognize faces, leading to social embarrassment and a sense of isolation. She felt a tumult of emotion, like someone exonerated from a crime. It seemed to explain so much about who she was.

To CONDUCT HIS FIRST INTERVIEW with Bill Choisser, Duchaine had to drive his red 1988 Camry 300 miles north to San Francisco. It wasn't easy for him. He'd been playing too much *V8*, a car combat videogame, and driving on a real freeway now gave him panic attacks. He'd pull over to calm himself down. It took a long time to get there.

When he did arrive, Choisser was clearly uncomfortable. Duchaine asked if everything was OK, and Choisser said he couldn't read Duchaine's expressions. It would help if he had a beard—facial hair made faces easier to comprehend.

So three weeks before each of his next eight road trips to San Francisco, Duchaine stopped shaving. He also cut back on playing *V8*, decreasing his freeway anxiety. He'd show up at Choisser's door with a scruffy beard and a newfound confidence, ready to run his subject through face-recognition tests, some of which he'd just invented. Once back in Santa Barbara, he would run the tests on dozens of grad students and faculty to generate averages for a representative sample. When the numbers came in, it was clear that Choisser's facial-recognition system was severely impaired.

Next, Duchaine tested Choisser on his ability to recognize small differences between the same type of objects. In one exam, Choisser was asked to memorize the details of a particular house. Duchaine then showed him 150 pictures of other houses and randomly threw in images of the original. Choisser consistently identified it. He did the same in similar tests that used shoes, horses, cars, and natural land-scapes. It was strong evidence that the brain's system for processing faces is separate from its system for discerning other objects.

The finding was controversial. "People love to carve up the brain and say, 'There's a hand area, there's a face area, and there's a love area,'" says Michael Tarr, a Brown University vision researcher who is one of the leading proponents of an alternative theory. Tarr believes that the fusiform face area is part of a generalized recognition system that gets switched on when someone develops enough expertise to distinguish between two very similar things.

"The discovery of otherwise healthy people living with prosopagnosia is the most exciting development in the field in a long time," Tarr says. "But we can't rely on neuropsychology to figure out what's going on, because you're just asking people questions. You're not getting in there and finding the damage."

The problem is that scientists aren't able to cut open someone's head, plug in electrodes, and test every neuron in the fusiform face area. But as Duchaine's work progressed, fMRI testing indicated that developmental prosopagnosics—those, like Choisser, who were born with the condition—had a smaller face area than normal. One study found that the neural pathways leading out of the face area might be damaged. But none of the tests explained the structure of the face-processing system itself. The best that could be done was to develop visual tests that narrowed down the nature of the impairment and then try to deduce what was going on inside the brain.

Using Choisser's group, now hosted on Yahoo, as a resource, Duchaine contacted and tested other developmental prosopagnosics and began piling up evidence that, with a bit of practice, they could distinguish between very similar objects. In other words, they could learn how to differentiate objects but not faces. The implication: Facial recognition was a hardwired aptitude that did not depend on learning.

As he found more prosopagnosics through Yahoo, Duchaine began to see that this previously unknown population offered a new way of exploring the brain's face-processing system. First of all, the patients were healthy and young, unlike gunshot or stroke victims. Their condition was stable, and the unobstructed blood flow in their brains allowed for clear imaging. They could also be studied over time. Duchaine had discovered an untapped well of subjects, and he was sure that a key to understanding how our brains are organized lay inside their heads.

For Ken Nakayama, a leading vision researcher at Harvard, Duchaine's dissertation on Choisser was a turning point. He had heard of developmental prosopagnosia—he had even seen it mentioned in the academic literature. But Duchaine's work with Choisser was so thorough and persuasive, it convinced Nakayama that the condition merited serious attention. "Brad did more than anybody else had done, by far," he says. "As soon as I saw his dissertation, I knew that this was what my lab needed to be studying."

Nakayama quickly offered Duchaine a job, and together they formed the Harvard Prosopagnosia Research Center. Though Duchaine still wore jeans, sneakers, and faded T-shirts to work, he was now a rising star in cognitive neuroscience. It suited him well— he stopped playing videogames, and the panic attacks subsided. In his free time, he pored over research papers on vision. If he stayed up late, it was to correspond with prosopagnosics in other parts of the world and to think about new ways to figure out what was going on in their heads. When particularly conclusive test results came in from a new subject, he would exclaim, "That guy's *really* bad with faces." It gave him a thrill. It was as if he'd spent his whole life waiting to stumble across this condition. Now that he'd found it, he had a prestigious job and a stable income. In exchange for shining a light on the disorder, prosopagnosia had made him an adult.

But research on face blindness was still in its infancy. Duchaine wanted to find out how our brains process faces and how that influences who we are. He and others hypothesize that one part of the face-processing system might be designed specifically for processing gender, another for skin attributes, and a third for reading emotional expression. To test the theory, he had to find double dissociations: a subject who could look at a face and perceive, say, skin color but not gender, and a second subject who had the opposite condition. That would provide evidence that skin color and gender perception were separate brain mechanisms.

The key was to find many different combinations of deficits. It would require lots of subjects, which is exactly what the discovery of developmental prosopagnosia promised. Before that, a scientist might have waited a lifetime to come across a stroke victim who had just one set of problems. Now, thanks to Choisser and the Internet, Duchaine could locate hundreds—even thousands—of subjects. "A whole new world of questions can now be asked," he says. "We're moving inside the black box."

Duchaine's questions touch on some fundamental social and cultural issues. For instance, what would it mean if there were a particular part of the brain devoted to recognizing gender? Bill Choisser reports that he has more trouble perceiving women's faces, and that could be one of the reasons he's gay. Another prosopagnosic says that his inability to distinguish between men and women explains his bisexuality. Is it possible that our sexuality is influenced by the wiring in the face-processing system?

Tom Uglow, a graphic designer in London, didn't have a problem perceiving that it was a girl watching him across the bar. Her blond hair had a nice sheen. She seemed pretty. Uglow ordered another beer, downed it, and walked her way. He was about to introduce himself when she cut him off.

"Hi, Tom," she said, no longer smiling. "Why were you making eyes at me?"

"Damn," he thought. "This isn't going as planned."

Her voice sounded familiar. He searched her face but couldn't place her. This happened more than he liked to admit.

"How've you been?" he asked, casually trying to fish for a clue as to who she was.

"Better now that we're broken up."

Ah! It was his ex-girlfriend. Once he'd had a moment to process her voice, he was able to place her. They had dated for a year. Definitely not a good person to be hitting on. It was a problem: Every time he saw a face, it felt like it was for the first time.

Uglow tried to work around it. Everyone looked equally unfamiliar, but rather than treat unfamiliar faces as strangers, he acted like everyone was his best pal. "Generally, I'll be very smiley, friendly, and nice, even though I have no idea who I'm talking to," he says. "But at least that person would come away liking me." He'd rather live in a world populated with friends than with strangers.

It never occurred to Uglow that his inability to see the world as others did would stand in the way of becoming an artist. Ever since he was a kid, he had loved drawing and made endless sketches of friends and family in his journals. They almost never had faces. For Tom, that wasn't a big deal. He felt he could convey personality with his brushstrokes or a particularly adept representation of someone's posture.

Every year, he made a pilgrimage to London's National Portrait Gallery. The museum hosts an annual contest to select the best emerging portrait artists, and Uglow likes to keep up with the latest techniques. This year, the winning portrait was a photorealistic depiction of an elderly woman, her face wrinkled and blotchy. Uglow loved it, though not because of the three-quarters-crazy look in the woman's eyes or her haunted expression. He loved it because the technique was so precise, so exact. To him, it spoke of the struggle against chaos, decay, and death. It was an attempt to impose order on what couldn't be controlled. In his own way, he had grasped the essence of the painting.

By day, Uglow makes a living designing logos, but someday he'd like to see his portraits hang in the gallery. The fact that they don't have faces doesn't seem to be an impediment. "Faces aren't that important in the contemporary art world," he observes. And yet, despite the missing faces—or maybe because of them—his paintings and sketches are evocative. In one sketch, a young girl holds an umbrella in a field. The umbrella has a densely patterned green and purple canopy, and individual blades of grass are visible. The girl's scarf is sharply defined, but she has no face. In another drawing, newlyweds stand beside a wedding cake; the bride has no head, but the fringe of her dress is elaborately delineated. It's a view into another reality.

In 2004, UNIVERSITY COLLEGE in London wooed Duchaine with the offer to run his own lab. He took the opportunity but continued to collaborate with Harvard's Nakayama. In 2005, the two conducted a study to determine the extent of prosopagnosia in the general population. They assessed 1,600 people online, running them through a face-recognition test the researchers had invented, and found that 32 had severely impaired face recognition. At the same time, a German researcher tested 680 high school and college students and identified 17 prosopagnosics. Both studies suggested a prevalence of roughly 2 percent. If the ratio held, it would mean that nearly 6 million people in the US are face blind.

Duchaine says the implications are far-reaching. He believes, for example, that the Transportation Safety Administration should administer tests to make sure all airport passenger screeners can match faces with IDs. And the reliability of eyewitness identification should be reconsidered in the courtroom. "You'd want to know if the witness was drunk, right?" Duchaine says. "Well, we should also know if they're face blind."

Baseline testing should be done at a young age so prosopagnosic children don't develop feelings of isolation and depression. There is no cure, but as with dyslexia or autism, once parents know a kid has the condition, they can make accommodations. It can be hard emotionally, particularly if a child doesn't recognize his parents. But it's worse if parents chastise a kid for something that can't be controlled. And, for the children of prosopagnosics, it can be comforting to know why their parent can sometimes seem oblivious to them.

MORDECHAI HOUSMAN, a genial, portly Hasidic Jew, is playing *Minesweeper* on his computer at home in Brooklyn. One of his three young sons sits next to him—though Housman isn't sure which.

"Who are you?" Housman finally asks with a smile.

"I'm Abraham, Dad," Abraham says. The 6-year-old has heard this question before and thinks his father is just kidding. It's like a family joke. He doesn't understand that his dad really can't tell him apart from other kids on the street.

The Hasidic men in the neighborhood—clad uniformly in black pants, white shirt, black jacket, and black hat—are even harder to distinguish. But this doesn't bother Housman. He considers it a blessing of sorts. "In my culture," he says, "we de-emphasize material things and appearances. Our focus should be on God. For me, I've been given a head start."

Nonetheless, even the religious have to make a living. Housman found work as a swing-shift building manager for a Hasidic college in Borough Park, Brooklyn. Most nights, he sits behind a small window near the entrance, answers questions, and is supposed to keep an eye on who is coming and going.

Housman doesn't think he's ill-suited for the task; he has developed other ways to recognize people. On a recent Monday night, he peers out of his small window, chews on a pen while beads of sweat gather on his forehead, and watches people walk past. He recognizes one student by his "intense, inwardly focused attitude." He greets a professor he thinks he recognizes by the type of yarmulke he is wearing, though it turns out to be the wrong guy. A young man darts past, but Housman knows him. "I recognize him by his soul," he says.

For Housman, as for most prosopagnosics, the Internet is changing everything. He learned about his condition when his wife stumbled across Choisser's Web site one night. She thought it explained a lot about her husband, and she showed the site to him. Housman stayed up until five in the morning reading Choisser's online memoir and following links to other prosopagnosia Web sites. For Hasidics,

Housman says, "it's considered arrogant if you make a big deal out of something or call attention to yourself. So I didn't really talk about my problems until I sent my first e-mail to Brad Duchaine."

For the past five years, Duchaine has been hungrily collecting profiles of developmental prosopagnosics though his Web site, Faceblind.org. It's a war chest assembled to search for double dissociations. There is the Caucasian man who upset a colleague by telling him a racial joke, not recognizing that the listener was black. There is the woman who waved to someone in a hotel, only to discover that it was a mirror. There is the man who takes emotional cues from the way a person's pants crease and bunch.

Now Duchaine is trying to prove that the condition is inherited. He plans to spend this New Year's in Las Vegas at a reunion of a family full of prosopagnosics. Three generations will be there. By sequencing DNA samples from every attendee and looking at the differences between those who do and don't have the condition, Duchaine hopes to pinpoint a particular gene or set of genes that code for face perception. It's unlikely to lead to a cure, but it could illuminate how the brain builds specialized abilities.

Developmental prosopagnosia came to light in large part because of Internet groups. Before that, most people born with the condition assumed they were just bad with faces. It's not the type of thing most would go to a doctor about, and even if they did, their physician probably couldn't help, because many doctors are unaware of it. In many ways, this is a neurological condition discovered by Yahoo.

Which makes Duchaine wonder if other groups of people with perception problems will start to coalesce online. It would certainly help him if new groups formed with names like Trouble Recognizing Gender or Trouble Recognizing Myself. Among the millions of Internet users, there are sure to be some who consider themselves normal save for one troubling quirk. They may hold the key to a deeper understanding of the way we assemble reality.

OLIVER SACKS

Stereo Sue

FROM *THE NEW YORKER*

It is easy for us to take our ability to see in three dimensions for granted. Meeting a woman who only late in life developed stereoptic vision, however, forced the renowned neurologist Oliver Sacks to ponder the complexities of this remarkable faculty.

WHEN GALEN, in the second century, and Leonardo, thirteen centuries later, observed that the images received by the two eyes were slightly different, neither of them appreciated the full significance of these differences. It was not until the early eighteen-thirties that the English scientist and inventor Charles Wheatstone began to suspect that the disparities between the two retinal images were in fact crucial to the brain's mysterious ability to generate a sensation of depth—and that the brain somehow fused these images automatically and unconsciously.

Wheatstone confirmed the truth of his conjecture by an experimental method as simple as it was brilliant. He made pairs of drawings of a solid object as seen from the slightly different perspectives of the two eyes, and then designed an instrument that used mirrors to insure that each eye saw only its own drawing. He called it a stereoscope, from the Greek for "solid vision." If one looked into the stereoscope, the two flat drawings would fuse to produce a single three-dimensional drawing poised in space.

(One does not need a stereoscope to see stereo depth; it is relatively easy for most people to learn how to "freefuse" such drawings, simply by diverging or converging the eyes. So it is strange that stereopsis was not discovered centuries before: Euclid or Archimedes could have drawn stereo diagrams in the sand, as David Hubel has remarked, and discovered stereopsis in the third century B.C. But they did not, as far as we know.)

A few years after Wheatstone's discovery came the invention of photography, and stereophotographs, with their magical illusion of depth, became immensely popular. Queen Victoria herself was presented with a stereoscope after admiring one at the 1851 Great Exhibition, at the Crystal Palace, and soon no Victorian drawing room was complete without one. With the development of smaller, cheaper stereoscopes, easier photographic printing, and even stereo parlors, there were few people in Europe or America who did not have access to stereo viewers by the end of the nineteenth century.

With stereophotographs, viewers could see the monuments of Paris and London, or great sights of nature like Niagara Falls or the Alps, in all their majesty and depth—with an uncanny verisimilitude that made them feel as if they were hovering over the actual scenes. (By the mid-eighteen-fifties, a subspecialty of stereophotography, stereopornography, was already well established, though this was of a rather static type, because the photographic processes used at the time required lengthy exposures.)

In 1861, Oliver Wendell Holmes (who invented the popular handheld Holmes Stereo Viewer), in one of several *Atlantic Monthly* articles on stereoscopes, remarked on the special pleasure people seemed to derive from this magical illusion of depth:

> The shutting out of surrounding objects, and the concentration of the whole attention . . . produces a dream-like exaltation . . . in which we seem to leave the body behind us and sail into one strange scene after another, like disembodied spirits.

There are, of course, many other ways of judging depth: occlusion of distant objects by closer objects, perspective (the fact that distant objects appear smaller), shading (which delineates the shape of objects), "aerial" perspective (the blurring and bluing of more distant objects by the intervening air), and, most important, motion parallax—the change of spatial relationships as we move our heads. All these cues, acting in tandem, can give a vivid sense of reality and space and depth. But the only way to actually *perceive* depth rather than judge it is with binocular stereoscopy.

In my boyhood home, in London, during the nineteen-thirties, we had two stereoscopes: a large, old-fashioned wooden one, which took glass slides, and a smaller, handheld one, which took cardboard stereophotographs. We also had books of bicolor anaglyphs—stereophotographs printed in different colors, which had to be viewed with a pair of red-and-green glasses that effectively restricted each eye to seeing only one of the images. So when I developed a passion for photography, at the age of ten, I wanted, of course, to make my own pairs of stereophotos. This was easy to do, by moving the camera horizontally about two and a half inches between exposures, mimicking the distance between the two eyes. (I did not yet have a double-lens stereocamera, which would take simultaneous stereo pairs.)

I started taking pictures with greater and greater separations between them, and then, using a cardboard tube about a yard long and four little mirrors, I made a hyperstereoscope—turning myself, in effect, into a creature with eyes a yard apart. With this, I could look at even a very distant object, like the dome of St. Paul's Cathedral, which normally appeared as a flat semicircle on the horizon, and see it in its full rotundity, projecting toward me. I also experimented with the opposite of this by making a pseudoscope (another device invented by Wheatstone), which transposed the views of the two eyes. This reversed the stereo effect to some extent, making distant objects appear closer than near ones, and turning faces into hollow masks, even though this contradicted common sense, and contradicted all the other depth cues of perspective and occlusion—a bizarre and disorienting experience as the brain struggled to reconcile two opposite conclusions, and alternated rapidly between rival hypotheses.

After the Second World War, new techniques and forms of stereoscopy became popular.

The View-Master, a little stereoscope made of plastic, took reels of tiny Kodachrome transparencies that one flicked through by pressing a lever. I fell in love with faraway America at this time, partly through such View-Master reels of the grand scenery of the West and the Southwest.

One could also get Polaroid Vectographs, in which the stereo images were polarized at right angles to one another; these were viewed through a special pair of Polaroid glasses, with the polarization of the lenses also at right angles, insuring that each eye saw only its own image. Such Vectographs, unlike the red-and-green anaglyphs, could be in full color, which gave them a special appeal.

Then there were lenticular stereograms, in which the two images were printed in alternating narrow vertical bands covered by clear ridged plastic. The ridges served to transmit each set of images to the proper eye, eliminating the need for any special glasses. I first saw a lenticular stereogram just after the war, in the London Tube— an advertisement, as it happened, for Maidenform bras. I wrote to Maidenform, asking if I could have one of their advertisements, but got no reply; they must have imagined I was a sex-obsessed teenager, rather than a simple stereophile.

Finally, there was a slew of 3-D films (like the Madame Tussaud's horror film, *House of Wax*), which one would look at through red-and-green or Polaroid glasses. As cinema, most of these were awful—but some, like *Inferno*, were very beautiful, and used stereophotography in an exquisite, delicate, unintrusive way.

Over the years, I amassed a collection of stereograms and books about stereoscopy.

I am an active member of the New York Stereoscopic Society, and at their meetings I encounter other stereo buffs. Unlike most, we do not take stereoscopy for granted but revel in it. While most people may not notice any great change if they close one eye, we stereophiles are sharply conscious of the change, as our world suddenly loses its spaciousness and depth. Perhaps we rely more on stereoscopy, or perhaps we are simply more aware of it. We want to understand how it works. The problem is not a trivial one, for if one can understand stereoscopy, one can understand not only a simple and brilliant visual stratagem but something of the nature of visual awareness, and of consciousness itself.

SOME PEOPLE, losing binocular vision for a long period, find the experience very disturbing. In a recent issue of *Binocular Vision & Strabismus Quarterly*, Paul Romano, a sixty-eight-year-old retired pediatric ophthalmologist, recounted his own story of losing nearly all sight in one eye, following a massive ocular hemorrhage. After one day of monocular vision, he noted, "I see items but I often don't recognize them: I have lost my physical localization memory. . . . My office is a mess. . . . Now that I have been reduced to a two-dimensional world I don't know where anything is."

The next day, he wrote, "Things are not the same object at all monocularly as they were binocularly. . . . Cutting meat on the plate—it is difficult to see fat and gristle that you want to cut away. . . . I just don't recognize it as fat and gristle when it only has two dimensions."

After almost a month, though Dr. Romano was becoming less clumsy, he still had a sense of great loss:

> Although driving at normal speed replaces the loss of depth perception with motion stereopsis, I have lost my spatial orientation. There is no longer the feeling I used to have of knowing exactly where I am in space and the world. North was over here before— now I don't know where it is. . . . I am sure my dead reckoning is gone.

His conclusion, after thirty-five days, was that "even though I adapt better to monocularity every day, I can't see spending the rest of my life in this way. . . . Binocular stereoscopic depth perception is not just a visual phenomenon. It is a way of life. . . . Life in a two-dimensional world is very different from that in a three-dimensional world and very inferior." As the weeks passed, Dr. Romano became more at home in his monocular world, but it was with enormous relief that, after nine months, he finally recovered his stereo vision.

In the nineteen-seventies, I had my own experience with losing stereoscopy when, following surgery for a ruptured quadriceps, I was put in a tiny windowless room in a London hospital. The room was scarcely bigger than a prison cell, and visitors complained of it, but I soon accommodated, and even enjoyed it. The effects of its limited horizon did not become apparent to me until later, as I described in *A Leg to Stand On*:

> I was moved into a new room, a new spacious room, after twenty days in my tiny cell. I was settling myself, with delight, when I suddenly noticed something most strange. Everything close to me had its proper solidity, spaciousness, depth—but everything farther away was totally flat. Beyond my open door was the door of the ward opposite; beyond this a patient seated

in a wheelchair; beyond him, on the windowsill, a vase of flowers; and beyond this, over the road, the gabled windows of the
house opposite—and all this, two hundred feet perhaps . . .
seemed to lie like a giant Kodachrome in the air, exquisitely colored and detailed, but perfectly flat.

I had never realized that stereoscopy and spatial judgment could
be so changed after a mere three weeks in a small space. My own
stereoscopy had returned, jerkily, after about two hours, but I wondered what happened to prisoners, confined for much longer periods. I had heard stories of people living in rain forests so dense that
their far point was only six or seven feet away. If they were taken out
of the forest, it was said, they might have so little idea or perception
of space and distance beyond a few feet that they would try to touch
distant mountaintops with their outstretched hands.

WHEN I WAS A NEUROLOGY RESIDENT, in the early nineteen-
sixties, I read the papers of David Hubel and Torsten Wiesel, who
later received a Nobel Prize for their work. They revolutionized our
understanding of how mammals learn to see, in particular of how
early visual experience is critical for the development of special cells
or mechanisms in the brain needed for normal vision. Among these
are the binocular cells in the visual cortex that are necessary to construct a sense of depth from retinal disparities. Hubel and Wiesel
showed that if normal binocular vision was rendered impossible by
a congenital condition (as in Siamese cats, often born cross-eyed) or
by experiment (cutting one of the muscles to the eyeballs, so that the
subjects became wall-eyed), these binocular cells would fail to
develop, and the animals would permanently lack stereoscopy. A
significant number of people are born with similar conditions—
collectively known as strabismus, or squint—a misalignment sometimes too subtle to attract notice but sufficient to interfere with the
development of stereo vision.

Yet there are many accounts of stereo-blind people who achieve remarkable feats of visuomotor coördination. Wiley Post, the first person to fly solo around the world, as famous in the nineteen-thirties as Charles Lindbergh, did so after losing an eye in his mid-twenties (he went on to become a pioneer of high-altitude flight, and invented a pressurized flight suit). A number of professional athletes have been blind in one eye, and so was at least one eminent ophthalmic surgeon.

There are many others—perhaps five or ten per cent of the population—who, for one reason or another, have little or no stereo vision, though often they are not aware of this, and may learn it only after careful examination by an ophthalmologist or optometrist. They may not all be pilots or world-class athletes, but many of them have no sense of visual impairment, either. Most manage to get along very well using only monocular cues, though some do have difficulty judging depth, threading needles, or playing sports.

There may even be certain advantages to monocular vision, as when photographers and cinematographers deliberately renounce their binocularity and stereoscopy by confining themselves to a one-eye, one-lens view, the better to frame and compose their pictures. And those who have never had stereopsis but manage well without it may be hard put to understand why anyone should pay much attention to it. Errol Morris, the filmmaker, was born with strabismus, and subsequently lost almost all the vision in one eye, but feels he gets along perfectly well. "I see things in 3-D," he said. "I move my head when I need to—parallax is enough. I don't see the world as a plane." He joked that he considered stereopsis no more than a "gimmick" and found my interest in it "bizarre."

I tried to argue with him, to expatiate on the special character and beauty of stereopsis. But one cannot convey to the stereo-blind what stereopsis is *like;* the subjective quality, the quale, of stereopsis is unique and no less remarkable than that of color. However brilliantly a person with monocular vision may function, he or she is, in this one sense, totally lacking.

And stereopsis, as a biological strategy, is crucial to a diverse array of animals. Predators, in general, have forwardfacing eyes, with much overlap of the two visual fields and, presumably, stereoscopic vision; prey animals, by contrast, tend to have eyes at the sides of their heads, which gives them panoramic vision, helping them spot danger even if it comes from behind. An astonishing strategy is found in cuttlefish, whose wide-set eyes normally permit a large degree of panoramic vision but can be rotated forward by a special muscular mechanism when the animal is about to attack, giving it the binocular vision it needs for shooting out its tentacles with deadly aim.

In primates like ourselves, forwardfacing eyes have other functions. The huge, close-set eyes of many types of lemurs serve to clarify the complexity of dark, dense close-up foliage, which, if the head is kept still, is almost impossible to sort out without stereoscopic vision—and in a jungle full of illusion and deceit, stereopsis is indispensable in breaking camouflage. On the more exuberant side, aerial acrobats like gibbons might find it very difficult to swing from branch to branch without the special powers conferred by stereoscopy. A one-eyed gibbon might not fare too well—and the same might be true of a one-eyed lemur or cuttlefish.

Stereoscopy is highly conserved in such animals, despite its costs—the sacrifice of panoramic vision, the need for special neural and muscular mechanisms for coördinating and aligning the eyes, and, not least, for special brain mechanisms to compute depth from the disparities of the two visual images. Thus, in nature, stereoscopy is anything but a gimmick, even if some human beings manage, and may even do better, without it.

IN DECEMBER OF 2004, I received an unexpected letter from a woman named Sue Barry. She reminded me how we had met, in 1996, at a shuttle-launch party in Cape Canaveral (her husband, Dan, was then an astronaut). We had been talking about different ways of experiencing the world—how, for example, Dan and other astronauts lost their sense of "up" and "down" in the near-zerogravity conditions of outer space and had to find ways of adapting. Sue had then told me of her own visual world: she had been born cross-eyed, and so viewed the world with one eye at a time, her eyes rapidly and unconsciously alternating. I had asked if this was any disadvantage to her. No, she said, she got along perfectly well—she drove a car, she could play softball, she could do whatever anyone else could. She might not be able to see depth directly, as other people could, but she could judge it as well as anybody, using other cues.

I had asked Sue if she could *imagine* what the world would look like if viewed stereoscopically. Sue said she thought she could—after all, she was a professor of neurobiology, and she had read plenty of papers on visual processing, binocular vision, and stereopsis. She felt this knowledge had given her some special insight into what she was missing—she knew what stereopsis must be like, even if she had never experienced it.

But now, nine years after our initial conversation, she felt compelled to write to me about this question:

> You asked me if I could imagine what the world would look like when viewed with two eyes. I told you that I thought I could. . . . But I was wrong.

She went on to give me details of her visual history, starting with her parents noticing that she was cross-eyed a few months after she was born.

> The doctors told them that I would probably outgrow the condition. This may have been the best advice at the time. The year was 1954, eleven years before David Hubel and Torsten Wiesel published their pivotal papers on visual development, critical periods, and cross-eyed kittens. Today, a surgeon would realign the eyes of a cross-eyed child during the "critical period" . . . in

order to preserve binocular vision and stereopsis. Binocular vision depends on good alignment between the two eyes. The general dogma states that the eyes must be realigned in the first year or two. If surgery is performed later than that, the brain will have already rewired itself in a way that prevents binocular vision.

Sue did have operations to correct her strabismus, first on the muscles of the right eye, when she was two, and then of the left eye, and finally of both eyes, when she was seven. When she was nine, her surgeon told her that she could now "do anything a person with normal vision could do except fly an airplane." (Wiley Post, apparently, had already been forgotten by the nineteen-sixties.)

She no longer looked cross-eyed to a casual observer, but she was half aware that her eyes were still not working together, that there was still something amiss, though she could not specify what it was. "No one mentioned to me that I lacked binocular vision, and I remained happily ignorant of the fact until I was a junior in college," she wrote. Then she took a course in neurophysiology:

The professor described the development of the visual cortex, ocular dominance columns, monocular and binocular vision, and experiments done on kittens reared with artificial strabismus. He mentioned that these cats probably lacked binocular vision and stereopsis. I was completely floored. I had no idea that there was a way of seeing the world that I lacked.

After her initial astonishment, Sue began to investigate her own stereo vision:

I went to the library and struggled through the scientific papers. I tried every stereo vision test that I could find and flunked them all. I even learned that one was supposed to see a three-dimensional image through the View-Master, the toy stereo viewer that I had been given after my third operation. I found the old toy in my parents' home, but could not see a three-dimensional image with it. Everyone else who tried the toy could.

At this point, Sue wondered whether there might be any therapy by which she could acquire binocular vision, but "the doctors told me that it would be a waste of my time and money to attempt vision therapy. It was simply too late. I could only have developed binocular vision if my eyes had been properly aligned by age two. Since I had read Hubel and Wiesel's work on visual development and early critical periods, I accepted their advice."

TWENTY-FIVE YEARS PASSED—years in which Sue married and raised a family while pursuing an academic career in neurobiology. She had not tried to fly an airplane, but she had found she could do almost everything else, with her other, monocular ways of judging space and distance. Occasionally, she enjoyed showing off these special abilities:

I took some tennis lessons with an accomplished pro. One day, I asked him to wear an eye patch so that he had to hit the ball using only one eye. I hit a ball to him high in the air and watched this superb athlete miss the ball entirely. Frustrated, he ripped off the eye patch and threw it away. I am ashamed to admit it, but I enjoyed watching him flounder, a sort of revenge against all two-eyed athletes.

But when Sue was in her late forties new problems began:

It became increasingly difficult to see things at a distance. Not only did my eye muscles fatigue more quickly, but the world appeared to shimmer when I looked in the distance. It was hard to focus on the letters on street signs or distinguish whether a person was walking toward or away from me. . . . At the same time, my glasses, used for distance vision, made me far-sighted. In the classroom, I could not read my lecture notes and see the students at the same time. . . . I decided it was time to get bifocals or progressive lenses. I was determined to find an eye doctor who would give me both progressive lenses to improve my visual acuity and eye exercises to strengthen my eye muscles.

She consulted Dr. Theresa Ruggiero, an optometrist, who found that Sue's eyes were developing various forms of imbalance—this sometimes happens after surgery for strabismus—and that the reasonable vision she had enjoyed for decades was now being undermined.

Dr. Ruggiero confirmed that I saw the world monocularly. I only used two eyes together when looking within two inches of my face. She told me that I consistently misjudged the location of objects when viewing them solely with my left eye. She explained that the shimmering, the difficulty in focusing on distant objects, resulted from binocular rivalry. I was constantly switching eyes. Most importantly, she discovered that my two eyes were misaligned vertically. The visual field of my left eye was about three degrees above that of my right. Dr. Ruggiero placed a prism in front of my right lens that shifted the entire visual field of the right eye upward. . . . Without the prism, I had trouble reading the eye chart on a computer screen across the room because the letters appeared to shimmer. With the prism, the shimmer was greatly reduced.

("Shimmer," Sue later explained, was perhaps too mild a term, for it was not like the shimmer one might see with a heat haze on a summer day—it was, rather, a rapid alternation of the misaligned images from each eye, so that whatever she was seeing seemed to jump up and down eight or ten times a second, in a dizzying way.)

Sue got her new eyeglasses, complete with the prism, on February 12, 2002. Two days later, she had her first vision-therapy session with Dr. Ruggiero—a long session in which, using Polaroid glasses to allow a different image to be presented to each eye, she attempted to fuse the two pictures. At first, she did not understand what "fusion" meant how it was possible to bring the two images together—but after trying for several minutes she found she was able to do this, though only for a second at a time. Although she was looking at a pair of stereo images, she had no perception of depth—but nevertheless she had made the first step, achieving "flat fusion," as Dr. Ruggiero called it.

Sue wondered whether, if she could hold the eyes aligned for longer, this would allow not just flat fusion but stereo fusion, too. Dr. Ruggiero gave her further exercises to stabilize her tracking and hold her gaze, and she worked on these exercises diligently at home. Three days later, something odd occurred:

> I noticed today that the light fixture that hangs down from our kitchen ceiling looks different. It seems to occupy some space between myself and the ceiling. The edges are also more rounded. It's a subtle effect but noticeable.

In her second session with Dr. Ruggiero, on February 21st, Sue repeated the Polaroid exercise and tried a new one, using colored beads at different distances on a string. This exercise, known as the Brock string, taught Sue to fixate both eyes on the same point in space, so that her visual system would not suppress the images from one eye or the other but would fuse them together. The effect of this session was immediate:

> I went back to my car and happened to glance at the steering wheel. It had "popped out" from the dashboard. I closed one

eye, then the other, then looked with both eyes again, and the steering wheel looked different. I decided that the light from the setting sun was playing tricks on me and drove home. But the next day I got up, did the eye exercises, and got into the car to drive to work. When I looked at the rear-view mirror, it had popped out from the windshield.

Her new vision was "absolutely delightful," Sue wrote. "I had no idea what I had been missing." As she put it, "Ordinary things looked extraordinary. Light fixtures floated and water faucets stuck way out into space." But it was "also a bit confusing. I don't know how far one object should 'pop out' in front of another for a given distance between the two objects. . . . It is . . . a bit like I am in a fun house or high on drugs. I keep staring at things. . . . The world really does look different." She included some excerpts from her diary:

February 22: I noticed the edge of the open door to my office seemed to stick out toward me. Now, I always knew that the door was sticking out toward me when it was open because of the shape of the door, perspective and other monocular cues, but I had never seen it in depth. It made me do a double take and look at it with one eye and then the other in order to convince myself that it looked different. It was definitely out there.

When I was eating lunch, I looked down at my fork over the bowl of rice and the fork was poised in the air in front of the bowl. There was space between the fork and the bowl. I had never seen that before. . . . I kept looking at a grape poised at the edge of my fork. I could see it in depth.

March 1: Today, I was walking by the complete horse skeleton in the basement of the building where I work, when I saw the horse's skull sticking out so much, that I actually jumped back and cried out.

March 4: While I was running this morning with the dog, I noticed that the bushes looked different. Every leaf seemed to stand out in its own little 3-D space. The leaves didn't just overlap with each other as I used to see them. I could see the SPACE between the leaves. The same is true for twigs on trees, pebbles on the road, stones in a stone wall. Everything has more texture.

Sue's letter continued in this lyrical vein, describing experiences utterly novel for her, beyond anything she could have imagined or inferred before. She had discovered for herself that there is no substitute for experience, that there is an unbridgeable gulf between what Bertrand Russell called "knowledge by description" and actual "knowledge by acquaintance," and no way of going from one to the other.

One would think that the sudden appearance of an entirely new quality of sensation or perception might be confusing or frightening, but Sue seemed to adapt to her new world with remarkable ease. She was startled and disoriented at first, but for the most part she felt entirely, and increasingly, at home with stereoscopy. Though she continues to be conscious of the novelty of stereo vision, and indeed rejoices in it, she also feels now that it is "natural"—that she is seeing the world as it really is, as it should be. Flowers, she says, seem "intensely real, inflated," where they were "flat" before.

Sue's acquisition of stereoscopy after almost half a century of being stereoblind has been a constant source of delight, and a great practical benefit. Driving is easier, threading a needle, too, and when she looks down into her binocular microscope at work she can see paramecia swimming at different levels, and see this directly, rather than inferring it by refocussing the microscope up or down.

At seminars . . . my attention is completely captivated by the way an empty chair displays itself in space, and a whole row of empty chairs occupies my attention for minutes. I would like to take a whole day just to walk around and LOOK. I did escape today for an hour to the college greenhouse just to look at the plants and flowers from all angles.

Most of the phone calls and letters I receive are about mishaps, problems, losses of various sorts. Sue's letter, though, was a story not of loss and lamentation but of the sudden gaining of a new sense and sensibility, and, with this, a sense of delight and jubilation. Yet her letter also sounded a note of bewilderment and reservation: she did not know of any experience or story like her own, and was perplexed to find, in all that she had read, that the achievement of stereoscopy in adult life was "impossible." Had she always had binocular cells in her visual cortex, she wondered, just waiting for the right input? Was it possible that the critical period in early life was less critical than generally thought? What did I make of all this?

I MULLED OVER SUE'S LETTER for a few days, and discussed it with several colleagues, including Bob Wasserman, an ophthalmologist, and Ralph Siegel, a vision physiologist. A few weeks later, in February of 2005, the three of us went to see Sue at her home in Massachusetts, bringing along ophthalmological equipment and various stereoscopes and stereograms.

Sue welcomed us and, as we chatted, showed us some childhood photos, since we were interested in trying to reconstruct her early visual history. Her childhood strabismus, prior to surgery, was quite clear in the photographs. Had she *ever* been able to see in three dimensions, we asked? Sue thought for a moment, and answered yes, perhaps—very occasionally, as a child, lying in the grass, she might suddenly see a blade of grass stand out from its background. The grass would have to be very close to her eyes, within inches, to do this, and the standing out would last just a second or two—she had almost forgotten about this until we quizzed her. So there was a suggestion (if her memory was not playing her false) that Sue may have had a few brief and rare stereo experiences in early life, but there was no way to be certain of this.

Sue had written, in her letter, "I think, all my life, I have desired to see things in greater depth, even before I knew I had poor depth perception." Was it possible that the intensity of this wish had made her believe that she was seeing in stereo when she actually was not? It was important to test her with special stereograms that had no cues or clues as to depth—no perspective or occlusion, for example. I had brought one stereogram with lines of print—unrelated words and short phrases—that, if viewed stereoscopically, appeared to be on seven different planes of depth, but, if viewed with one eye, or without true stereo vision, appeared to be on the same plane. Sue looked at this picture through the stereoscope and saw it as a flat plane. It was only when I prompted her by telling her that some of the print was at different levels that she looked again, and said, "Oh, now I see." After this, she was able to distinguish all seven levels and put them in the correct order.

Given enough time, Sue might have been able to see all seven levels on her own, but such "top-down" factors—knowing or remembering or having an idea of what one should see—are crucial in many aspects of perception. A special attention, a special searching, may be necessary to reinforce a relatively weak physiological faculty. It seems likely that such factors are strongly operative with Sue, especially in this type of test situation. Her difficulties in real life are much less, because every other factor here—knowledge, context, and expectation no less than perspective, occlusion, and motion parallax—helps her experience the three-dimensional reality around her.

Sue was able to see depth in the red-and-green drawings I had brought. One of these images—an impossible three-pronged tuning fork such as M. C. Escher might have drawn, with three tines of increasing heights—Sue found "spectacular"; she saw the top of the uppermost prong as three or four centimetres above the plane of the paper. Bob and Ralph, by comparison, saw it as twelve centimetres above, and I saw it as fifteen centimetres above.

I found this surprising, because we were all the same distance from the drawing, and I had imagined that a given disparity would be perceived, invariantly, as a constant depth. Puzzled by this, I wrote to several neuroscientists, including Shinsuke Shimojo, at Caltech, an expert in many aspects of visual perception. He brought out, in his reply, that when one looks at a stereogram the computational process in the brain is based not solely on the binocular cue of disparity but also on monocular cues such as size, occlusion, and motion parallax. With a stereoscopic illusion, these cues conflict, the monocular ones working against the binocular ones. The brain must therefore balance one set of cues against the other, and arrive at a weighted average. This final result will be different in different individuals, because there is huge variation, even in the normal population: some people rely predominantly on binocular cues, others on monocular cues, and still others use both. In looking at a stereo illusion such as the tuning fork, a strongly binocular person will see unusual stereo depth; a monocularly oriented person will see much less depth; and others, relying more equally on both binocular and monocular cues, will see something in between.

LATER IN THE DAY, we paid a visit to Sue's optometrist, Dr. Theresa Ruggiero, who described how Sue had first consulted her, in 2001. Sue had complained then of eyestrain, especially when driving, impaired clarity, and a disconcerting jumping or flickering of images—but had not mentioned her lack of stereoscopy.

Dr. Ruggiero herself was greatly surprised, she said, when, immediately after achieving flat fusion, Sue experienced stereoscopy. She speculated that Sue must have had some binocular vision and stereoscopy, even if very briefly, during the critical childhood period, or it would not have been possible for her to have stereo vision now. What was so remarkable about Sue, Ruggiero said, over and above the initial achievement of stereoscopy, was her adventurous and positive reaction to it, and her fierce determination to hold on to it and enhance it, however much work this might entail.

And it did indeed entail, and still entails, a great deal of work—taxing fusion exercises for at least twenty minutes every day. With these exercises, Sue found that she was starting to perceive depth at greater and greater distances, where at first she had seen depth only close up, as with the steering wheel. She continued to have jumps of improvement in her stereo acuity, so that she was able to see depth with smaller and smaller disparities—but when she stopped therapy for six months she quickly regressed. This upset her deeply, and she resumed the eye exercises by herself, working on them every day. Three years later, she still does them, "religiously."

Sue has continued to work very hard on her stereo perception and stereo acuity in the months since our visit, and her perception of stereo depth has continued to increase measurably. Moreover, she has developed a skill she did not have when we initially visited her: the ability to see random dot stereograms. Unlike conventional stereo pictures, these are constellations of dots with no images that can be seen monocularly, but which reveal images or shapes when viewed with both eyes. This illusion may take some practice, and many people, even with normal binocular vision, are not able to get it. But often, as one continues to gaze, a strange sort of turbulence appears among the dots, and then a startling illusion—an image, a shape, whatever—will suddenly appear far above, or far below, the plane of the paper. Getting these illusions is the purest test of stereoscopic vision. It is unfakable, for there are no monocular cues whatever; it is only by stereoscopically fusing thousands of seemingly random points as seen by the two different eyes that the brain can construct a three-dimensional image.

Though a theoretical understanding of random-dot stereograms came only in the nineteen-sixties, they are akin to the stereo illusions described by David Brewster, the inventor of another early stereo viewer, as early as 1844. Gazing at wallpaper with small repetitive motifs, he observed that the patterns might quiver or shift, and then jump into startling stereoscopic relief, especially if these patterns were offset in relation to one another. Such "autostereograms" have probably been experienced for millennia, with the repetitive patterns of Islamic art, Celtic art, the art of many other cultures. Medieval manuscripts such as the Book of Kells or the Lindisfarne Gospels, for example, contain exquisitely intricate designs done so exactly that whole pages can be seen, with the unaided eye, as stereoscopic illusions. (John Cisne, a paleobiologist at Cornell, has suggested that such stereograms may have been "something of a trade secret among the educated élite of the seventh and eighth century British Isles.")

In the past decade and a half, elaborate autostereograms have been widely published as *Magic Eye* books. These have added another dimension to Sue's newfound stereoscopic powers: "I find these wallpaper autostereograms easy (and quite thrilling)," she recently wrote, "probably because I practice convergent and divergent fusion regularly."

IN THE SUMMER OF 2005, Bob and I paid Sue another visit, in Woods Hole, Massachusetts, where she was running a fellowship program in neurobiology. She had mentioned to me that the bay there was sometimes full of luminous organisms, mostly tiny dinoflagellates, and that she enjoyed swimming among them. When we arrived, in the middle of August, we found that our timing was perfect—the water was aflame with the luminous creatures ("*Noctiluca scintillans*—I love the name," said Sue). After dark, we went down to the beach, armed with masks and snorkels. We could see the water sparkling from the shore, as if fireflies were in it, and when we immersed ourselves and moved our arms and legs in the water, clouds of miniature fireworks lit up around our limbs. When we swam, the night lights rushed past our eyes like the stars streaking past the *Enterprise* as it reaches warp speed. In one area, where the noctiluca were particularly dense, Bob said, "It's like swimming into a galaxy, a globular cluster."

Sue, overhearing this, said, "Now I see them in 3-D—they all seemed to twinkle in a flat plane before." Here there were no contours, no boundaries, no large objects to occlude or give perspective.

There was no context whatsoever—it was like being immersed in a giant random-dot stereogram, and yet Sue now saw the noctiluca at different depths and distances, in three-dimensional space. If she could do this, we mused, perhaps she could now do even better on the random-dot stereogram tests. But Sue, normally eager to talk about stereo vision, was mesmerized by the beauty of the scintillating organisms. "Enough thinking!" she said. "Give yourself to the noctiluca."

STRUGGLING TO FIND AN ANALOGY for her experience, Sue had suggested, in her original letter to me, that her experience might be akin to that of someone born totally color-blind, able to see only in shades of gray, who is suddenly given the ability to see in full color. Such a person, she wrote, "would probably be overwhelmed by the beauty of the world. Could they stop looking?" While I liked the poetry of Sue's analogy, I disagreed with the thought, for I suspect that someone who has grown up in a completely colorless world would find it confusing, or even impossible, to integrate a new "sense" such as color with an already complete visual world. Color, for such a person, would have no associations, no meaning.

But Sue's experience of stereoscopy was clearly not a gratuitous or meaningless addition to her visual world. After a brief confusion, she embraced the new experience, and felt it not as an arbitrary add-on but as an enrichment, a natural and delicious deepening of her existing vision. Perhaps this was because a three-dimensional world was already a perceptual reality for her, even though she had relied on non-stereoscopic means to achieve it. With color, there is no precursor—we either see a world of color or we do not—but all of us live and move in a three-dimensional world.

David Hubel has followed Sue's case with interest, and has corresponded with her and with me about it. He has pointed out that we are still quite ignorant of the cellular basis of stereoscopy. We do not know, even in animals, whether disparity-sensitive cells (the binocular cells specialized for stereoscopy) are present at birth (though Hubel suspects they are); what happens to these cells if there is strabismus and lack of binocular experience in early life; and, most crucially, whether they can recover if the strabismus is repaired. With regard to Sue, he wrote, "It seems to me that [her regaining of stereopsis] occurred too quickly for it to be due to a reestablishment of connections, and I rather would guess that the apparatus was there all along, and just required reestablishment of fusion to be brought out." But, he added, "that's just a guess!"

Whether the cells and mechanisms that enable binocular stereoscopy are present at birth or form soon thereafter, the notion of a critical period of maximum sensitivity to environmental stimuli still stands fast. Without early binocular experiences, these cellular mechanisms either die out or fail to develop. But if there is any binocular vision at all during the critical period—and with strabismus there may still be some overlap of the visual fields of each eye and thus a small area of fusion—then the essential apparatus for stereopsis may be established.

What emerges from Sue's experience is that there seems to be sufficient plasticity in the adult brain for these binocular cells and circuits, if some have survived the critical period, to be reactivated later. In such a situation, though a person may have had little or no stereo vision that she can remember, the potential for stereopsis is nonetheless present and may spring to life—most unexpectedly—if good alignment of the eyes can be obtained. That this seems to have happened with Sue after a dormant period of almost fifty years is very striking.

Though Sue originally thought her own case unique, she has found, on the Internet, accounts by a number of other people with strabismus and related problems who have unexpectedly achieved stereo vision through vision therapy.

And a report that has just been published in the journal *Nature* described the case of S.K., a twenty-nine-year-old man who was born without lenses in his eyes. Though functionally blind for his entire life (he could sense little more than light and dark), he was able to acquire competent vision after being given a pair of glasses. Such an acquisition, long after the critical period, would traditionally have been considered extremely improbable. But S.K.'s case, like Sue's, suggests that if there are even small islands of function in the visual cortex, there may be a fair chance of reactivating and expanding them in later life, even after a lapse of decades, if vision can be made optically possible. Cases like these may offer new hope for those once considered incorrigibly blind or stereo-blind.

WHATEVER ITS NEUROLOGICAL BASIS, the augmentation of Sue's visual world has effectively granted her an added sense, a circumstance that the rest of us can scarcely imagine. For her, stereopsis continues to have a quality of revelation.

"After almost three years," she wrote, "my new vision continues to surprise and delight me. One winter day, I was racing from the classroom to the deli for a quick lunch. After taking only a few steps from the classroom building, I stopped short. The snow was falling lazily around me in large, wet flakes. I could see the space between each flake, and all the flakes together produced a beautiful three-dimensional dance. In the past, the snow would have appeared to fall in a flat sheet in one plane slightly in front of me. I would have felt like I was looking in on the snowfall. But now, I felt myself within the snowfall, among the snowflakes. Lunch forgotten, I watched the snow fall for several minutes, and, as I watched, I was overcome with a deep sense of joy. A snowfall can be quite beautiful—especially when you see it for the first time."

STACEY BURLING

Probing a Mind for a Cure

FROM THE *PHILADELPHIA INQUIRER*

In reporting on how scientists try to understand the cause of—and find a cure for—Alzheimer's disease and other forms of dementia, Stacey Burling also shows the impact of this ravaging disease on a once-vital man and his anguished family.

BOB MOORE'S BRAIN lay on a white plastic cutting board. There was something beautiful about its convoluted hills and valleys, the way rivers of dusky purple and red meandered through the beige flesh.

And mysterious. Here was the essence of a man who had gone to Yale, loved a woman, fathered six children, relished ice cream and Mozart and Kierkegaard and e.e. cummings, favored questions over answers and change over complacency, hated camping, loathed golf, and, over the last 20 years, had slowly lost the capacity to understand any of it.

He had died that morning in a Wilmington nursing home, years past being able to feed himself or walk or recognize the woman he had married 56 years before.

What had gone wrong with his brain?

Before neuropathologist Mark Forman lifted his knife last December in a basement autopsy suite at the Hospital of the University of Pennsylvania, he could see that Bob Moore's brain wasn't normal. But it would be weeks before he could tell Moore's family what had made the man they loved disappear long before his heart stopped beating.

Robert B. Moore, a Presbyterian minister, was a spiritual man, but he was also a believer in science and medicine.

After being diagnosed with Alzheimer's disease in 1993, at age 67, he entered a clinical trial of an experimental drug. He let doctors, intent on finding ways to detect dementia earlier, tap his spinal fluid and compare it with healthy people's.

And he decided that his brain would be autopsied at Penn's Center for Neurodegenerative Disease Research, founded and run by two nationally prominent dementia researchers.

Doctors can tell with about 90 percent accuracy whether a patient has Alzheimer's, the most common dementia. But looking through a microscope at brain tissue after death is still the only way to diagnose it with absolute certainty.

Perfecting diagnosis is critical in the emerging era of drugs designed for specific types of dementia.

But diagnosis is just the beginning. By studying brains from patients such as Bob Moore, scientists hope to figure out how and why the damage occurred—and learn to prevent it. More than one in five women and one in six men who reach age 65 will develop dementia before they die, a study this month reported. By 2050, more than 13 million Americans will have Alzheimer's, another study estimated.

"We want to cure this damn disease," said John Trojanowski, a neuropathologist who launched the center in 1982 with his wife, Virginia Lee, a biochemist and cell biologist. The center performed 86 autopsies last year—its record. Researchers there publish 40 to 50 scientific papers a year.

For Trojanowski, this is an epic battle.

"I talk about this as a threat to our economy and way of life that equals any natural disaster, and I think it will be worse. . . . The baby boomers start turning 60 this year," said Trojanowski, who is 59 himself. "We really are in a race with time."

For Moore's family, it's personal.

What exactly, they wonder, turned this gentle man into an unruly stranger who shouted obscenities and hit his wife? Why did the disease strike him a decade earlier than the average? Could the car accident that gave him amnesia briefly in 1950 have triggered this calamity years later?

Because of their family history, Moore's children also feel some urgency about the science that their father's brain cells may fuel. Three of their grandparents also died with dementia. Even their 78-year-old mother, who jokes that she is a "normal control" in Penn's studies, has a gene that raises risk for Alzheimer's.

With most now well into middle age, the children—a pediatric physical therapist, an architect, a computer consultant, a classical musician, a foreign-service officer, and a daughter whose own ministerial career was sidetracked by an autoimmune disease—all worry about their futures.

At 51, Alison Moore, the would-be minister, already has taken medicine for mild cognitive impairment, a condition that often precedes dementia.

Betsy Shieh, 45, the foreign-service officer, said her memory doesn't seem any worse than her peers' now. Still, she says, "when you forget something when your father's just died of Alzheimer's, you say: 'Oh s—, here I go.'"

Long after it was obvious that he was losing his fight with dementia, Bob Moore believed that doctors might save him. His children hope the cure he was so optimistic about will come in time for them.

———

BOB MOORE CAME INTO THE WORLD with a superior brain. He used it hard and well.

As a young man, his fascination with life's big questions led him to Yale University's Divinity School and, later, a master's in philosophy. It drew him to the civil rights movement and a lifelong quest to broaden his church and improve the lot of the poor.

In 1970, he rose to executive of the Delaware-based New Castle Presbytery, the denomination's regional ruling body.

A 1978 sermon, in which he mused about the meaning of faith, revealed a complex man and graceful writer, a minister who saw religion not as an easy comfort, but as a struggle for self- and world-betterment. He quoted the Bible, of course, but also the *New York Times*, Dostoyevsky, and Thomas Hobbes.

"I believe that faith is like the free, innocent, life-entrusting leap of a child into the arms of a parent," he said. That trust erodes as adults learn that "faith does not always, does not often, win."

Ultimately, he said, "our leap of faith is a lifetime thing, and what is unknown remains unknown for us and what is hidden remains hidden. Faith is the posture of our lives, the bent of our spirit, the foundation and definition of who we are."

An only child, the robust, red-haired man who liked quiet and order had three sons and three daughters with Joanna, a warm, practical preacher's daughter he met in college. He was the animated, emotional one—a belly-laugher and great hugger. She was stoic, gentle, forgiving. They fit.

Although he would sometimes call for silence as he listened to his classical music, he mostly delighted in the chaos of their Wilmington home. He was a joyful, supportive father.

"When my dad was proud of you, you knew it," said Alison Moore, who followed him to Yale Divinity School. "His whole body would light up . . .

"You felt like you were illuminated."

———

"THIS IS A REALLY SMALL BRAIN," Forman said on Dec. 1 as he took his first look at case number 05-274.

A normal man's brain weighs 1,200 to 1,400 grams. Bob Moore's weighed 1,005 grams—a little more than 2 pounds. The folds at its surface were half the usual thickness, a sign that many cells had died.

Several diseases with similar symptoms can cause this kind of atrophy. A surprising number of patients have more than one. Knowing whether they had Alzheimer's, frontotemporal dementia, or dementia with Lewy bodies is important for families because each carries genetic risk. It helps doctors sharpen diagnoses in preparation for the day when they'll have different treatments for different dementias. Twenty-seven dementia drugs are now in human clinical trials or awaiting approval.

Forman looked at and felt the surface of the brain. It was symmetrical, as it should be. No areas were squishier than normal. There were yellow flecks in the carotid arteries, evidence of blood-vessel clogging. The olfactory bulbs were exceptionally small. Poor sense of smell is an early sign of dementia, and Moore did, in fact, do poorly on a scent test soon after his diagnosis.

His brain stem was also unusually small—not typical of Alzheimer's patients, Forman pointed out, but not unheard of.

He cut the brain stem from the rest of the brain and looked at the substantia nigra, dark lines that run through the midbrain. In Parkinson's disease, these lines lighten. The lines in Moore's brain had lost some pigment, but less than usual for Parkinson's.

Using a long knife, Forman sliced the gelatin-like brain in half. The center would use one hemisphere for the autopsy. The other would be frozen and used only for research.

Again, Forman, a wiry, graying man whose grandmother had Alzheimer's, noticed atrophy. The ventricles—pools of cerebrospinal fluid—were much larger than normal, a sign that surrounding tissue had died.

Forman cut the hemispheres into half-inch slices. Each had brown scalloped edges—the gray matter or nerve cells—and creamy, white centers—the electrical wiring. He saw no signs of tumors or stroke.

The big-picture work over, Forman took samples from about 20 parts of the brain, including the hippocampus, the seat of memory and the place where Alzheimer's is thought to start; the amygdala, an emotional center that's also involved in smell; the brain stem, responsible for involuntary activities such as heartbeat and breathing. Especially shrunken was the visual cortex, a part of the brain usually unaffected by Alzheimer's.

What Forman saw was "consistent" with Alzheimer's, but, he said, "none of this tells me definitively what the diagnosis is going to be. Every patient doesn't read the book." He still had to study Moore's brain cells.

Forman's caution would prove justified.

BOB MOORE BEGAN COMPLAINING about his memory to his doctor in 1985. He wasn't even 60 yet, and the doctor discounted his fears. Most people don't have noticeable problems until their early- to mid-70s.

By 1988, Bob Criswell, pastor of the church that Moore attended, was also concerned. He pulled Moore aside to tell him he was repeating himself.

"Does it show that much?" Moore asked.

A few months later, his assistant at the presbytery, Bob Bolt, gently told Moore that he was returning phone calls more than once.

Within days, Moore, who'd had some health problems and was feeling the stress of his job, announced his retirement. He was 62.

Still curious and energetic, he began teaching philosophy at the University of Delaware's Academy of Lifelong Learning. But late in 1992, after he struggled with an ambitious class on philosophy and religion, he questioned his students. They said he'd been making mistakes.

Moore demanded an appointment with a specialist. A psychiatrist told him that he probably had Alzheimer's.

"How long do I have?" Moore asked.

"About 10 years," the psychiatrist said.

With characteristic forthrightness, Moore told each of his children. In a letter to his son Tom, now a computer consultant in New York, he wrote: "I'm in good shape psychologically. I'm not depressed . . . However, I will not teach anymore. I know I could, but I also know that I have sometimes been embarrassed in classes I have taught when I have made minor slips, not just once but several times. . . . I am sensitive about this. . . .

"All people die. . . . I have a good chance of living many years yet, with very gradually decreased acuity."

Bob Moore did, indeed, live well for several years. He took classes and enjoyed them. He and Joanna traveled.

In a 1996 letter to their children, Joanna Moore, a retired teacher who thought it important to record their family history, described the disease's progress. Her husband could still drive well, with her help as navigator, but could not do simple arithmetic. He still enjoyed reading and going to movies. He had trouble with time and was unable to follow through on plans. In conversation, he failed at finding even ordinary words. He could no longer compare two thoughts or visualize something.

"He wants to open his own mail, answer the phone, continue the routines he has followed for many years," Joanna wrote. "Bob does the dishes and usually does the laundry. I'm grateful. He washes the kitchen floor and vacuums the rugs. He locks the house at night and turns out the lights and pulls the blinds. . . .

"Normal life consists of a hundred small responsibilities, and I hope we can continue our normal life for a long, long time."

FROM AUTOPSIES OF PEOPLE who died in different stages of Alzheimer's, scientists know that its first physical signs usually appear in the medial temporal lobes. These reception areas for sensory input near the temples help produce memories.

Particularly hard hit is the hippocampus, a three-inch-long portion shaped like a sea horse that records new memories. Trojanowski calls it "the epicenter of disaster."

From there, the protein hallmarks of Alzheimer's—plaques, or clumps of protein fragments that form outside nerve cells, and tangles, twisted strands of a different protein that form inside cells—move up, forward and back through the brain. Scientists do not know how or why.

The amygdala, which is beside the hippocampus, is an early casualty. Damage there may cause emotional changes. As the disease moves along the sides of the brain, it affects language skills. In the frontal lobe near the top of the brain, executive function—the ability to make rational decisions—is disrupted. As this section erodes, people find it harder to plan and understand time. At roughly the same time, the parietal lobe, which helps people orient themselves in space, is damaged, making it harder for people to find their way around.

The path and its connection to behavior grow hazier after that because the disease destroys not only brain cells but also the connections between them, isolating regions that may still look fine. "It just sits there and rips the brain apart," Trojanowski said.

There's no clear physical explanation for the incessant walking, agitation and violent behavior that some patients exhibit.

Some agitation likely stems from the fear that constant surprises breed. As memories fade, people do not know how to react.

"Their frame of reference is disappearing," said Christopher Clark, director of Penn's Memory Disorders Clinic and Bob Moore's doctor after his diagnosis. "You know who you are based on your past. You use that to project what's going to happen in the future. As your past disappears, your ability to project into the future essentially disappears, too."

Ultimately, Alzheimer's destroys enough of the brain that patients can't walk. By the time they can't swallow, the brain may be so far gone that it's hard to pinpoint why.

"At one point," Trojanowski said, "this is a train wreck, and there's so much damage it's hard to say what killed the passengers."

As the '90s waned, Bob Moore became increasingly dependent on Joanna and increasingly angry.

For one who had taken such pride in his verbal abilities, being incapable of finding the right words was humiliating, Alison Moore said. Unable to marshal an argument, he'd exclaim: "Damn it! I am . . . I am . . ." Finally, he'd remember. "I'm a man," he'd say emphatically.

As the dementia worsened, his frustration grew. "I am . . . ," he'd stammer. "I am . . ."

The sentence would hang unfinished, a reminder of what he'd lost.

By 2000, an exhausted Joanna Moore was keeping a journal. Pages and pages are devoted to the vagaries of her husband's behavior. Like the mother of a difficult toddler, she focused on his bowels and his tantrums, and was grateful for the good days when he was pliable and she could get a full night's sleep.

She had to watch him now so he wouldn't step in front of cars. He was a dangerous passenger who grabbed the steering wheel or tried to open his door while the car was moving. He shouted swear words and walked constantly, sweeping belongings off tables as he strode. He had hit his oldest son, Andy, a Wilmington architect, and he had hit Joanna.

"Yesterday was the day I reached my own personal limit," she wrote that November, "fighting him to get clothes off, keep him in the shower while I washed him, getting him dressed. I think we've gone back to another good day/bad day cycle, but the good days (today) aren't very good, and the bad days are horrible."

The next month, she decided she could no longer handle him. Rejected as too violent by nursing homes, Moore was admitted to a psychiatric hospital. To Joanna's horror, he rode there in handcuffs.

By spring 2001, she was no longer sure her husband remembered her or other family members. She had decided to decline hospital care if he got sick, choosing hospice, instead. "I know that this was Bob's choice, as well," she wrote. "He is now in the process of dying. I expect the disease will steadily deprive him of his humanity until there is nothing left but the shadow of life. First it took his memory; then it took his dignity; now it is taking his life."

A CENTURY AGO, when German scientist Alois Alzheimer first identified the disease that bears his name, doctors thought dementia was rare. In reality, few lived long enough to get it.

By the time Trojanowski and Lee decided to take on Alzheimer's, the dark side of a lengthening life span was emerging. The longer people's bodies lived, the more often their minds failed.

In 1982, scientists knew little about the diseases that make brains die. They did not even know what made up the plaques and tangles that Alzheimer had seen under his microscope.

Lee and Trojanowski would prove that tangles are made of an abnormal form of the protein tau. Later, they showed how it weakens and kills cells. Other scientists found that plaque contained a different protein, beta-amyloid. In 1998, the protein in Lewy bodies—round clumps found in one part of the brain in Parkinson's patients and in other parts in those with Lewy body dementia—was identified.

Trojanowski and Lee then proposed a unified theory of dementia. All of the dementias, they said, have in common misfolded proteins that accumulate in the brain. How to derail that process is the billion-dollar question.

Much remains a mystery.

How and why do cells die? What role do the errant clumps of protein play? What makes the diseases appear in more and more of the brain? How can some people have Alzheimer's pathology without symptoms? What triggers violence?

Then there's the key question for Moore's children: What's my risk? For late-onset Alzheimer's, which starts after age 60, there's no easy answer. Genes certainly are a factor, though only one for this type has been found.

One study estimated that 39 percent of people with a parent or sibling who'd had Alzheimer's will eventually develop the disease.

Another of male twins released this month found that genes mattered most, but lifestyle seemed to make a difference. Twins who did not have Alzheimer's when their brothers did tended to have more education. This supports the theory that "cognitive reserve" can help prevent or delay the disease, said the study's author, Margaret Gatz, a University of Southern California psychologist.

Experts say there's no way to predict an individual's late-onset Alzheimer's risk. It's even murkier for other dementias.

"I don't think it means anything to the kids that they can deal with one way or the other," Clark said of the Moore family history.

He could only suggest activities that seem to lower risk: Eat healthy food, exercise, and keep your mind active. Moore's children do all that.

It was January 2004, and Joanna arrived at Bob's nursing home to feed him lunch, just as she had almost every day for two years.

Bob was in a padded chair with a sloped back so he could slump without falling out. The legs of his gray sweatpants rode up almost to his knee, revealing his thin, pale shins. He made singsongy sounds that an administrator explained as singing and preaching. It helped other patients and families accept his outbursts. He wore thick glasses. Joanna didn't know whether he could see or whether he even knew she was there.

He had trouble swallowing, so he was fed ground or mashed food. Joanna gave him milk, rubbing his shoulder as she held a baby's sippy cup to his lips. Unable to cough properly, he cleared his throat with a loud, staccato laugh—"Ah hah hah hah"—that fell like a rollercoaster.

While waiting for his bean soup to arrive, Joanna looked at him intently. "Hi, sweetie," she said, tenderly brushing his wavy hair back from his forehead. On some days, she thought he responded to her touch. She'd whisper "I love you" in his ear, and he'd babble a response. Maybe some part of him still knew her. On this day, he sometimes looked right at her, but there was no hint of understanding.

He still loved to eat, and it was keeping him alive. Once a big man—5-foot-10 and 200 pounds—he had dropped to 130 at one point. Now, he was back up to 156. His health had improved so much that hospice was no longer appropriate.

"When I start blithering, I want you to shoot me," he had told Joanna after he got sick.

He was blithering now, but she couldn't follow his orders. Well-aware of the irony, she spooned the soup, a big helping of pureed chicken and noodles, mashed carrots, a roll, sugar cookies softened in milk, and ice cream into his gaping mouth.

When the food was gone, she wheeled him back to his room. She lowered the back of the chair until it was almost flat and gently removed his glasses. Not knowing whether it mattered, she turned on a classical music tape. In no time, he was asleep.

LAST FALL, Bob Moore's trouble with swallowing worsened. He got pneumonia, probably because improperly swallowed food lodged in his lungs. Told that he would die within days, Joanna called their children home.

On the morning of Nov. 30, the sons—Tom, Andy and John, a double-bass player who'd left a tour with the Pittsburgh Symphony Orchestra—sat uncomfortably around their father's bedside, reminiscing and staring at the pale, open-mouthed man whose chest rose and fell beside them.

"Feel free to touch him," Joanna told her sons. "I got a lot of comfort last night out of sitting here and just touching him." His warmth had soothed her.

"You just realize there's such a big break between life and death," she said.

That night, the daughters arrived—Kathy Thomas, the physical therapist from Pittsburgh; Betsy from Hyde Park, N.Y.; and Alison, of

Wilmington. They cried, surprised that their father's imminent death freed them to remember the man they had loved.

He breathed quietly in bed as his children's voices swirled around him.

The six of them and a grandson held hands with Joanna as they circled the bed to pray. "I am thankful for this unbroken chain of family," Joanna said.

Bob Moore died before the sun rose. His wife and children were relieved.

ON THE AFTERNOON OF FEB. 8, Forman opened the first of five folders containing 60 slides of Bob Moore's brain.

Each of the slides held a fragile slice of brain tissue treated with antibodies and chemicals to bring out, with color or light, the protein landmarks of dementia.

As always, Forman started with the fuchsia slides, the standard stain for seeing cell structure.

What struck him right away was what was not there. Slides from a normal brain would be solid pink. Many of these were dappled with white spots that made them look like slices of baby Swiss or leaves eaten down to the veins. The white holes were the abandoned homes of dead cells.

The hippocampus was sparsely populated, with far fewer pyramid-shape neurons than normal. Even with this stain, not designed to show clumps of proteins, Forman saw his first tangles—dark, flame-shape interlopers flowing from some cell nuclei.

This was typical for Alzheimer's.

But in the brain stem, he saw something intriguing: planet-like circles with pale rims and dark pink centers.

These were Lewy bodies. Forman would expect to find them in the brain stems of patients with Parkinson's, but with more damage. This might mean that Moore's disease was more complicated than his doctors had thought.

Forman moved on to the next slides. These had been stained to reveal abnormal tau, the key component in tangles, by turning it brown. Slides from a normal brain would be all white, with a bluish cast. Moore's had a dark ribbon of brown on the edges.

"I call that my Alzheimer's disease diagnosis," Forman said. "I don't even have to look at the slide to know what the disease will be." He looked anyway and saw what he had expected: "wall-to-wall" pathology.

Curious about what he'd found in the brain stem, Forman now picked up the set of slides that stained alpha-synuclein, the protein in Lewy bodies, dark brown.

Developed in 1998, the stain has turned up Lewy bodies in many Alzheimer's patients, raising questions about this protein conglomeration's relationship with plaques and tangles.

In half of Alzheimer's patients, Lewy bodies are in the amygdala—the part of the brain key to processing emotions. In a quarter, the Lewy bodies are widespread enough to warrant a different diagnosis.

Doctors often miss this dual problem while patients are alive, probably because symptoms of the two dementias are so similar. But Lewy body dementia is more likely to involve hallucinations, sleep disruptions, fluctuations in thinking ability—plus the tremors and rigidity of Parkinson's.

Moore had not had movement problems, but he may have had some of the others.

Forman peered through his microscope and saw that Moore's amygdala was packed with Lewy bodies. The surprise, though, was how many Forman saw throughout the brain, particularly in the upper regions.

"This," he said, "is an extraordinary case of Lewy body pathology."

Finally, it was time for fluorescent slides that light up both the plaques and tangles of Alzheimer's in glow-stick green. The plaques—puffs of protein stuck together like popcorn balls—are seen best in these slides.

In a normal brain, Forman would see only black. The slide of

Moore's hippocampus lit up like the Milky Way, chock full of oddly shaped fluorescent stars and twisting streaks.

"We know what the diagnosis is," he said.

Bob Moore had gotten a dementia double whammy. Separately, he had enough plaques and tangles—and enough Lewy bodies—to ruin his brain.

Brain 05-274 would have a new job now. It would join 900 others stacked on shelves in the lab, waiting to help scientists vanquish their killers.

CLARK, BOB MOORE'S DOCTOR, heard the news first and was surprised by how many Lewy bodies his patient had.

One of his research goals is to create simple tests of spinal fluid or urine that detect and sort out dementias early. He has already added Moore's information to his database. And, he said, the next time he sees an Alzheimer's patient with behavior problems like Moore's, he'll be more likely to think of Lewy bodies.

Joanna Moore slit open the envelope containing Forman's report on Tuesday.

She read it, re-read it, and read it again. She lingered on Forman's diagnosis: "Lewy body variant of Alzheimer's disease."

It was gratifying to see proof that Bob had, indeed, had Alzheimer's disease. But what was a Lewy body?

She still had questions, but having something concrete to explain what her family had been through felt good. Her children would want the news. "They're the inheritors," she said.

DAVID DOBBS

A Depression Switch?

FROM THE *NEW YORK TIMES MAGAZINE*

Despite great strides in treating depression over the past two decades, antidepressants and therapy help only about half of those affected. A new hypothesis suggests that the roots of depression lie in the brain's circuitry and not in its chemistry. David Dobbs reports on a new surgical technique that seems to flip the depression switch.

EANNA COLE-BENJAMIN NEVER FIGURED to be a test case for a radical new brain surgery for depression. Her youth contained no traumas; her adult life, as she describes it, was blessed. At 22 she joined Gary Benjamin, a career financial officer in the Canadian Army, in a marriage that brought her happiness and, in the 1990's, three children. They lived in a comfortable house in Kingston, a pleasant university town on Lake Ontario's north shore, and Deanna, a public-health nurse, loved her work. But in the last months of 2000, apropos of nothing—no life changes, no losses—she slid into a depression of extraordinary depth and duration.

"It began with a feeling of not really feeling as connected to things as usual," she told me one evening at the family's dining-room table. "Then it was like this wall fell around me. I felt sadder and sadder and then just numb."

Her doctor prescribed progressively stronger antidepressants, but they scarcely touched her. A couple of weeks before Christmas, she stopped going to work. The simplest acts—deciding what to wear, making breakfast—required immense will. Then one day, alone in the house after Gary had taken the kids to school and gone to work, she felt so desperate to escape her pain that she drove to her doctor's office and told him she didn't think she could go on anymore.

"He took one look," she told me later, "and said that he wanted me to stay right there in the office. Then he called Gary, and Gary came to the office, and he told us he wanted Gary to take me straight to the hospital."

They drove to the Providence Continuing Care Center's mental-health hospital, still known locally as the Kingston Psychiatric Hospital, or K.P.H., its name when it was built in the 1950's. "It's a dingy, archaic place," Deanna said, "typical of older mental hospitals." There, in the locked ward that also contained psychotic patients, she would spend the next 10 months straight and about 85 percent of the three years after that. Her depression would prove resistant to every class of antidepressant, numerous combinations of antidepressants and anti-anxiety drugs, intensive psychotherapy and about a hundred sessions of electroconvulsive therapy. Patients who have failed that many treatments usually don't emerge from their depressions.

Finally, in the spring of 2004, Deanna's psychiatrist at the hospital, Dr. Gebrehiwot Abraham, received a fax from a University of Toronto research team asking if he had an appropriate candidate for a clinical trial of a new, experimental surgery for treatment-resistant depression. The operation borrowed a procedure called deep brain stimulation, or D.B.S., which is used to treat Parkinson's. It involves planting electrodes in a region near the center of the brain called Area 25 and sending in a steady stream of low voltage from a pacemaker in the chest. One of the study's leaders, Dr. Helen Mayberg, a neurologist, had detected in depressed patients what she suspected was a crucial dysfunction in Area 25's activity. She hypothesized that the electrodes might modulate the area and ease the depression.

The procedure, Dr. Abraham told Deanna and Gary, had worked safely in thousands of Parkinson's patients. But it would carry some risk of neural complications (it was, after all, brain surgery), it would be uncomfortable and it might not work.

"We were in tears," said Deanna, who is now 41. "We felt we'd tried everything and nothing worked. But we talked about it and decided, 'Well, what have we got to lose?' "

What she hoped to lose, of course, was her depression. But depression, which 5 to 10 percent of Americans suffer in any given year and about 15 percent will suffer in their lifetimes, can be hard to lose. Drugs, as shown in a comprehensive study released last month by the National Institute of Mental Health, are effective in only half of patients with major depression. Psychotherapy does no better. For those people who are not helped by therapy or drugs, electroconvulsive therapy, or ECT, can bring relief. But few of those cures are complete. These therapies usually ease rather than cure depression while sometimes bringing side effects like insomnia or memory loss, and their potency often proves fleeting; as many as half to two-thirds of those successfully treated relapse within two years. Neither neuroscientists nor psychiatrists can say exactly what depression is. Neurologically and psychologically, what Hippocrates called the "black bile" and Susan Sontag "melancholy minus its charms" presents an almost impossibly complicated puzzle.

The expectations for the Toronto team's D.B.S. study were accordingly modest. When I later asked Mayberg's collaborator Dr. Andres Lozano, the neurosurgeon who performed the operations, what he had expected, he replied, "Nothing."

As it turned out, 8 of the 12 patients he operated on, including

Deanna, felt their depressions lift while suffering minimal side effects—an incredible rate of effectiveness in patients so immovably depressed. Nor did they just vaguely recover. Their scores on the Hamilton depression scale, a standard used to measure the severity of depression, fell from the soul-deadening high 20's to the single digits—essentially normal. They've re-engaged their families, resumed jobs and friendships, started businesses, taken up hobbies old and new, replanted dying gardens. They've regained the resilience that distinguishes the healthy from the depressed.

These results brought a marvelous surprise to both the patients and the doctors involved—and nervous anticipation about whether their luck will hold. Though a few of the patients are more than two years out from the surgery, none completely trust their cure. No one can tell them for sure that it will last, and they worry. The study doctors and the wider medical community, meanwhile, are guarded about whether D.B.S. will prove so effective in larger trials. "I can't emphasize enough that we need a large, randomized study to confirm this as a treatment," says Valerie Voon, a University of Toronto psychiatrist who was with the team for the first six patients and who is now a research fellow at the National Institutes of Health.

Those caveats notwithstanding, many scientists following the trial say they believe it will change how psychiatrists define and treat mood disorders. Mayberg, who speaks of a "paradigm shift," notes that she developed the trial to evaluate not a treatment but a hypothesis. In that sense the trial succeeded. Mayberg's focus on Area 25 tests the emerging "network" model of mood disorders, a new way of looking at psychiatric conditions that isn't restricted by the neurochemical model of mood that has dominated over the past quarter century or so. Rather, it incorporates neurochemistry into the concept of the brain as a circuit board or wiring diagram. The network model carries profound implications for research and, ultimately, treatment. The Prozac revolution showed everyone that tweaking neurochemistry can dampen and sometimes extinguish depression—but only through a generalized approach, hitting the entire brain. ("Carpet-bombing," one neuroscientist calls it.) And the 50 percent success rate of antidepressant drugs suggests that they aren't hitting depression's central mechanism. The network approach, on the other hand, focuses on specific nodes, pathways and gateways that might be approached with various treatments—electrical, surgical or pharmacological. This small trial appears to confirm this model so emphatically that it's already changing the neuropsychiatric view of the brain and the direction of research.

"People often ask me about the significance of small first studies like this," says Dr. Thomas Insel, who as director of the National Institute of Mental Health enjoys an unparalleled view of the discipline. "I usually tell them: 'Don't bother. We don't know enough.' But this is different. Here we know enough to say this is something significant. I really do believe this is the beginning of a new way of understanding depression."

When she started her research in the late 1980's, Helen Mayberg, too, looked at neurochemistry. "That's where biological psychiatry was then," she told me. "It was about the brain as a bowl of soup. You whip up a chemical, add it and stir. An alchemist point of view. But I soon realized I wanted to find out where things were changing."

Lively, smart and quick-witted, Mayberg talks of brain science with contagious excitement. She possesses just the kind of presence someone having a hole drilled in his head would welcome—authoritative but warm. Mayberg originally considered becoming a psychiatrist, but she didn't like the discipline's resistance at the time (it was the 1970's) to neurological explanations of mood. So she became a behavioral neurologist, doing a residency at Columbia University and then moving to Johns Hopkins.

Setting aside the bowl-of-soup model did not mean deciding that neurochemicals weren't important. Rather it meant deciding that neurochemistry, and particularly the chemistry dictating how individual neurons communicate with one another, was probably driven by traffic between different brain areas, and that identifying the patterns in that traffic might yield new understanding. (Or, using another metaphor, if the brain is an orchestra, then the neurochemical approach focuses on how well individual players listen and respond to the players adjacent to them; the network approach, like a conductor, focuses on how the orchestra's sections—strings, winds, brass, etc.—coordinate and balance volume and tone. When both are working well, you've got music.)

Imaging tools for tracking these relationships, like PET scans and later functional magnetic resonance imagery, were just maturing as Mayberg pursued her work. Neuroscientists were soon using these tools to help identify networks involved in mental processes ranging from distinguishing facial expressions to experiencing alarm or pleasure. Each of these networks engages different brain areas in different combinations. The areas active in recognizing a fearful expression, for instance, won't match those for recalling old memories, though some areas might overlap. Defining the network involved in any given process requires figuring out not just which parts are involved but also which parts are most vital and how one affects another.

By the 1990's, Mayberg was trying to define the network that goes awry in depression. She and other researchers soon established that depression involved abnormal patterns of activity in a network that includes limbic areas (a cluster of evolutionarily older brain areas around the top of the brain stem), which control basic emotions and drives like fear, lust and hunger, and the newer cortex and subcortex responsible for thought, memory, motivation and reward.

Several researchers were working on this. But Mayberg, and, separately, Dr. Wayne Drevets, then at Washington University and now at the N.I.M.H., increasingly homed in on Area 25, which seemed crucial in both its behavior and its position in this network. They found that Area 25 was smaller in most depressed patients; that it lighted up in every form of depression and also in nondepressed people who intentionally pondered sad things; that it dimmed when depression was successfully treated; and that it was heavily wired to brain areas modulating fear, learning, memory, sleep, libido, motivation, reward and other functions that went fritzy in the depressed. It seemed to be a sort of junction box, in short, whose malfunction might be "necessary and sufficient," as Mayberg put it, to turn the world dim. Maybe it could provide a switch that would brighten the dark.

To work the switch, Mayberg needed a knife. In 1999 she moved from the University of Texas at San Antonio to the University of Toronto, where she met Lozano, who had become expert at using deep brain stimulation for treating Parkinson's, the neurological affliction that causes tremors and rigidity as well as cognitive and emotional declines. By the time he and Mayberg met, he'd slipped electrodes into the brains of almost 300 patients.

Depression is more elusive than Parkinson's. But approaching Area 25 with D.B.S. allowed the researchers to use a known tool. Neurosurgeons found as early as the 1950's that they could treat Parkinson's by destroying a small portion of the hyperactive globus pallidus, a brain area that is crucial to movement. The treatment illustrated one of the brain's many oddities: some areas can cause more trouble when they are excessively active than when they have no activity at all. In the early 1990's, surgeons increasingly began to use D.B.S. to quiet the globus pallidus by sending it steady, rapid pulses of low voltage. Patients' tremors would instantly ease or cease; their rigidity and uncontrollable body movements would fade over a week or two. Killing the current revived the symptoms. Surgeons have now implanted D.B.S. electrodes in some 30,000 Parkinson's patients worldwide. The procedure is not a cure-all. It helps some patients less than others, does little to alleviate Parkinson's cognitive and emotional decay and occasionally creates complications including infection, bleeding and memory loss. Its biggest problem may be its success. So many medical centers now do it that some do it badly or on poorly qualified patients. But done well, it usually works.

Mayberg knew all this from the literature and learned more in conversations with Lozano. She grew increasingly convinced that applying D.B.S. to Area 25 might control depression.

"So one day," she told me, "I went to Lozano and said: 'I want to turn off Area 25. Can we put a stimulator there and see if that does it?' And he goes, 'Why not?'"

———

OCCASIONALLY, Deanna felt good enough to go home. This feeling seldom lasted more than a few days. "You could tell she was getting bad again when she couldn't sleep," Gary said. "That was the red flag. She'd be around the house all night, watching TV, up worried, cleaning. Then she'd get worse each day. Her eyes got that sunken look. Those were the scariest times, when she was getting like that and I would drop the kids at school and go to work and know she was home alone."

During the bad periods, which was much of the time, Deanna thought about suicide almost constantly. Through the windows of the locked ward she could see Lake Ontario, cold and immense. While she was there, one patient managed to reach the lake, beyond the parking lot and a grove of trees, and drown himself. Deanna thought obsessively of doing the same.

"I imagined that all the time," she said. "That I would walk out there and walk into the lake and that would be it."

As the months and years passed and all treatments failed, it began to feel as if there were only one way out.

"It started to seem like, this is not going to stop," Gary said. "This is our life now. There were times I thought that it was going to end"— he looked across the table at Deanna—"only when you committed suicide."

"The worst part for me," Deanna said, "was not being able to feel anything for my children. To hug them, to have them hug me, and feel nothing. That was devastating. An awful, awful place to be."

The D.B.S. operation involves an intrusion that is delicate but brutal. The patients are kept awake so they can describe any changes, and the only drug administered is a local anesthetic. The surgical team shaves much of the patient's head and attaches to the skull, with four screws drilled through skin into bone, the stereotactic frame that will hold the head steady against the operating table and serve as a navigational aid. Mounting the frame takes only about 10 minutes. But because it involves driving screws into the skull ("You can't truly feel it," as one patient said, "but you can hear it and see it and smell it"), and because it leaves you with a steel frame around the head, many patients find this the most distressing part of the whole business.

Gary found the frame more than he could take. He kissed his wife and went elsewhere, hoping she wouldn't be a vegetable when he next saw her. Then Deanna was rolled to an M.R.I. machine, where scans would be taken; the scans would help guide Lozano in placing the electrodes.

During the hour or so while the computer processed the scans, Deanna chatted with Mayberg. The day before, she told Mayberg, on video, that what she most wanted was to hold her children and feel it.

When the scans were ready, she went to the operating room. She was placed on a table, which was tilted back like a La-Z-Boy. Lozano and his team bolted the stereotactic frame to the table, as it was described to me later. There was some scrubbing on her head, some chitchat among the surgical team, much fiddling with sterile drapes and instruments. Then Lozano fit a half-inch burred bit into a drill, turned it on and started drilling. He drilled right into the top of Deanna's skull, which brought a rattling sensation and a sound like that made by an air wrench removing the lugs off a car's wheel. Then he did it again.

Now Lozano threaded a guide tube—"It's a straight shot," he said later, "really quite easy"—down between crevices and seams to one side of Area 25, which is in two small lobes at the midline of the brain. He slid the first electrode and its lead down the tube, then repeated this for the other side. All this took nearly two hours. After he double-checked his locations, he wired the leads to a pacemaker and gave Mayberg a nod. They could turn it on anytime now.

Mayberg had squeezed into a spot at Deanna's side some time before. She had told Deanna that if anything felt different, she should say so. Mayberg wasn't going to tell her when the device was activated. "Don't try to decide what's important," Mayberg told her. "If your nose itches, I want to know." Now and then the two would chat. But so far Deanna hadn't said much.

"So we turn it on," Mayberg told me later, "and all of a sudden she says to me, 'It's very strange,' she says, 'I know you've been with me in the operating room this whole time. I know you care about me. But it's not that. I don't know what you just did. But I'm looking at you, and it's like I just feel suddenly more connected to you.' "

Mayberg, stunned, signaled with her hand to the others, out of Deanna's view, to turn the stimulator off.

"And they turn it off," Mayberg said, "and she goes: 'God, it's just so odd. You just went away again. I guess it wasn't really anything.'

"It was subtle like a brick," Mayberg told me. "There's no reason for her to say that. Zero. And all through those tapes I have of her, every time she's in the clinic beforehand, she always talks about this disconnect, this closeness and sense of affiliation she misses, that was so agonizingly painful for her to lose. And there it was. It was back in an instant."

Deanna later described it in similar terms. "It was literally like a switch being turned on that had been held down for years," she said. "All of a sudden they hit the spot, and I feel so calm and so peaceful. It was overwhelming to be able to process emotion on somebody's face. I'd been numb to that for so long."

It worked that way for other patients too. For those for whom it worked, the first surges of mood and sensation were peculiar to their natures. Patient 4, for instance, was fond of taking walks, and she had previously told Mayberg that she knew she was getting ill when whole landscapes turned dim, as if "half the pixels went dark." Her first comment when the stimulator went on was to ask what they'd done to the lights, for everything seemed much brighter. Patient 5, an elite bicycle racer before his depression, told me that a pulling that he had long felt in his legs and gut, "as if death were pulling me downward," had instantly ceased. Patient 1, who in predepression days was an avid gardener, amazed the operating room by announcing that she suddenly felt as if she were walking through a field of wildflowers. Two days after going home, she put a scarf over her shaved, stitched head, found her tools and went out to reclaim her long-neglected gardens.

Not all was light and flowers. On a purely biological level, the improvement made by D.B.S. sometimes amplified the side effects of the high doses of medication the patients had been taking. Doctors don't quite understand this phenomenon, but they see it happen in other instances too; it is as if the patient, deadened, is again made sensate. Deanna broke out in hives and felt nauseated; her hands shook. These symptoms eased when she (as several of the patients have done) reduced her meds—slowly, so as not to introduce new variables. She now takes standard doses of Effexor, an antidepressant, and Seroquel, an anti-psychotic drug.

The cure also brought challenges at home. As with other disabilities, any partner turned caretaker gets used to calling the shots, and rearranging power, dependencies and expectations after a sudden recovery can prove hard. One patient, the cyclist, faced this challenge starkly, for he had started a relationship and married while he was depressed. "Frankly," he told me, "I'm not sure we've quite finished working this out." All the patients have benefited from coordinated assistance from psychiatrists, social workers and occupational therapists who try to smooth the transition.

"That help is crucial," says Mayberg, who is now a professor of psychiatry and neurology at Emory University in Atlanta. "We're just fixing the circuit. The patient's life still needs work. It's like fixing a knee. They need that high-quality physical and supportive therapy afterward if they're really going to move around again."

This transition is not back to a former self and family but to a new one. Gary Benjamin says he sees similar things in military families. "These soldiers get sent away for six months, they come back and all they want to do is return to their old home. But their old home isn't there, because everybody's changed. It takes some tough rearranging sometimes."

For a change so profound, these seem acceptable adjustments. And the treatment so far seems remarkably free of side effects. No one has suffered significant neural complications, probably because, unlike ECT, which sends 70 to 150 volts through the entire brain, these electrodes deliver only about 4 volts to an area about the size of a pea.

But what will happen when larger groups are treated? The team is continuing to operate on depressed patients, with a goal of 20. And would the successes stay high and the side effects low in a large, placebo-controlled trial? Neither Mayberg nor any of the other collaborators cares to guess. Other treatments have started this well and fizzled. For instance, vagus nerve stimulation, which sends a low current to the brain via a major nerve with connections to various brain areas, appeared to help about half of the patients in a small, initial, uncontrolled trial, but failed its only placebo-controlled trial. (In a controversial move, the Food and Drug Administration overruled its own reviewers and approved the device as a depression treatment anyway.)

"What if you do a hundred patients," I asked Mayberg one day, "and they do no better than placebo?"

"I suppose that's possible," she said. But she doesn't think that will be the case. The several authorities I talked to agree that the high success rate so far, along with the soundness of the theoretical base and D.B.S.'s track record with Parkinson's, suggests that this isn't just a lucky run.

"This just makes so much sense," says Dr. Antonio Damasio, director of the University of Southern California's Brain and Creativity Institute and a renowned neurologist, "and the weight of the results is so sizable. I would be surprised if they had no results with a larger body of patients."

On the other hand, even if it works, no one sees this becoming the new Prozac. The procedure costs too much (around $40,000) to use on anyone who hasn't tried everything else. The appropriate candidates for D.B.S. probably number in the thousands, not the millions. Perhaps the most sensible worry is that if the thing works, doctors might use it too freely, as they tend to do with successful new treatments; witness the problematic boom in D.B.S. for Parkinson's.

In the end, the procedure's greatest clinical value may lie in inspiring less intrusive ways of tweaking key nodes—localized delivery of drug or gene therapies, or other means still to come. Such possibilities probably lie at least a decade away.

Regardless of how it pans out in the clinic, Mayberg and Lozano's D.B.S. study is already changing how neuroscientists and psychiatrists think about depression. One possibility, for instance, is that refining the networks that go awry in depression may reveal neurological subtypes of depression that can be diagnosed and treated differently. For example, Mayberg has already found that patients who respond well to Prozac usually show a change in their brain scans only a week after they start medication—even though they don't feel a difference for 3 to 10 weeks (a long and sometimes dangerous wait). She's done preliminary work suggesting she might be able to identify such Prozac-friendly patients before they even start the drug. If she or others can replicate and elaborate on this diagnostic ability, doctors might be able to characterize a patient's depression and choose a best-odds therapy—Prozac for one patient, talk therapy for a second, both for a third—at the very beginning of treatment, saving weeks or months of trial and error.

The network model—which some scientists also call a "systems" model—also offers an organizing principle for research. Andreas Meyer-Lindenberg, a researcher with the N.I.M.H., points out that research on depression has so far followed clues left by drugs that were found to be effective only by chance. "We'd find a drug that helped depression, figure out how it works and make hypotheses from that about how the brain works," he says. The effectiveness of selective serotonin reuptake inhibitors (S.S.R.I.'s) like Prozac, for instance, inspired piles of research showing that mood regulation depends heavily on the availability of the neurotransmitter serotonin. (Neurotransmitters are chemicals that help carry messages across the spaces between neurons; S.S.R.I.'s treat depression by making more serotonin available in those spaces.)

This focus on neurotransmitters is the "bowl of soup" approach that Mayberg speaks of, and it has formed the bulk of depression research for more than two decades. Defining the networks the neurotransmitters move within, however—and in particular identifying

Area 25 as a key gateway within the depression network—will let researchers bring their neurochemical knowledge to bear on specific targets.

"With this D.B.S. work," Meyer-Lindenberg says, "they have characterized in detail a system"—or network—"underlying a major disorder. It's not a simplistic thing where you're saying it's all about this one area and you inject a current and everything's fine. It's a very complicated system. But this D.B.S. work shows us that amid this complicated system there is a place of overlap, a common denominator"—Area 25—"that's a very attractive treatment target." Here, Meyer-Lindenberg says, researchers can try to apply the knowledge they've gained about neurochemistry and genetics. The network theory presents a framework around which to apply these perspectives.

Meyer-Lindenberg's own work shows the power of this approach. Last June he published a study on the serotonin transporter gene, or sert gene, which helps determine serotonin availability. Other research had shown that people with the "short" version of the sert gene run more depression risk. Meyer-Lindenberg found a way to identify how various brain areas were affected by having that short version. Then he took 112 patients, half with the long version and half with the short, scanned their brains and asked the computer to find areas that scanned differently in the two groups. The area showing the most difference was Area 25.

Along with redirecting research, the quieting of Area 25 may also change our conception of depression from a condition in which something is lacking—self-esteem, resilience, optimism, energy, serotonin, you name it—to one in which an active agent makes a person sick.

"Most people think of depression as a deficit state," Mayberg says. "You're low, you're negative. But in fact, talk to a depressed person, and you have this bizarre combination of numbness and what William James called 'an active anguish.' 'A sort of psychical neuralgia,' he said, 'wholly unknown to healthy life.' You're numb but you hurt. You can't think, but you are in pain. Now, how does your psyche hurt? What a weird choice of words. But it's not an arbitrary choice. It's there. These people are feeling a particular, indescribable kind of pain."

This anguish, Mayberg suggests, is the manifestation of a neural circuit run amok. For doctors, establishing this should focus research and care. For those of us who've never known depression, recognizing it may help us see depression not as a dead absence but as a live affliction. We might even stop indulging the romantic notion of depression as intrinsic to one's identity. For this notion, too, was tested by Mayberg's experiment. When a steady, 4-volt thrum calmed these patients' anguish, they did not lose their identities. They regained them, feeling again the engagements with the world that most define them: flowers for the gardener, lightness for the cyclist and, for Deanna, a long-missed connection to others.

When Deanna, Gary and I finally finished talking, they insisted on driving me to my hotel. Halfway through town, Gary pulled off the main road, drove up a long, sinuous driveway and parked in a lot facing a dark, rambling building.

"This is the hospital," Gary said. "You see where Deanna stayed."

In the winter dark, the secure ward, off to the left, was easily discerned. It was a low wing, the only one with a few lights still on inside. Outside, bright flood lamps illuminated an exercise yard ringed by 20-foot-tall cyclone fencing topped with razor wire.

"And there's the lake," Gary said, motioning behind us. Through trees I could make out its blackness.

We sat several minutes, but no one said much.

"Well," Gary said, putting the minivan in gear. "We'd better get home."

DENISE GRADY

With Lasers and Daring, Doctors Race to Save a Young Man's Brain

FROM THE *NEW YORK TIMES*

A new laser device allows doctors to remove brain aneurysms, but has yet to be approved pending a determination of its safety. Denise Grady gets a glimpse of this new medical technology in action.

OPERATING ROOM 14, Dec. 12, 9:30 A.M.—"I always prep my own patients," Dr. David J. Langer said. "It relaxes me."

He picked up a sponge soaked in antiseptic and began scrubbing the shaved skull of Chris Ratuszny, 26, a mechanic from Lindenhurst, N.Y.

Mr. Ratuszny lay on the operating table, anesthetized and oblivious. His head jutted out past the end of the table, supported by four pins that had been screwed into his skull. The pins were attached, like spokes in a wheel, to a semicircular frame—surreal but standard, the hardware typically used to immobilize the head for brain surgery. A thick purple line had been drawn from his neck to the top of his head, to guide the scalpels.

He was about to become the first person in the United States to undergo an operation involving the use of an excimer laser to treat a giant brain aneurysm, a dangerous ballooning of an artery that could burst and kill him or leave him with devastating brain damage. The aneurysm was too big for the most common treatments, which involve clips or metal coils; it required bypass surgery on an artery in the brain.

The laser is not approved for brain surgery in the United States, but Dr. Langer got permission from the Food and Drug Administration to use it on an emergency basis for Mr. Ratuszny (ra-TOOSH-nee) last Tuesday at Roosevelt Hospital in Manhattan. The Dutch neurosurgeon who devised the laser procedure, Dr. Cornelius Tulleken, flew in from the Netherlands to help. He has performed the operation on about 300 patients in Europe.

Dr. Tulleken's technique involves a seemingly small variation on the standard procedure and takes just a few minutes in an eight-hour operation. But it could make all the difference for patients like Mr. Ratuszny, said Dr. Langer, who traveled to Utrecht in 1999 to learn the procedure from Dr. Tulleken. The advantage of the laser is that it lets surgeons operate without clamping a major artery in the brain—a step required in the standard operation, but one that can cause a stroke.

"It's a high-risk operation in the best of hands," Dr. Langer said.

He estimated that the laser could reduce the risk of stroke from bypass surgery for aneurysms to 12 percent, from 15 percent. But comparative studies have not been done. Some surgeons are skeptical, while others are eager to learn the technique, and it has begun to catch on in Europe, Dr. Tulleken said. A neurosurgeon from Chicago came to New York just to see how Mr. Ratuszny's procedure was done.

The laser definitely makes the operation easier, Dr. Langer said, because just knowing that the brain arteries are still open takes enormous time pressure off the surgeon during critical parts of the operation. To him, that alone makes it worthwhile.

"If it was me, my head, and there was a new device that would allow me to have this operation without occluding an artery, that's what I'd want," Dr. Langer said.

Besides making operations easier, the laser may make surgery possible for some aneurysms that would otherwise be inoperable, Dr. Tulleken and Dr. Langer say. Hoping to get the device approved in the United States, Dr. Langer plans to direct a study of it at several medical centers in the United States starting in March. The hospital invited *The New York Times* to observe and report on the operation, whatever the outcome. Even if the device is approved, it is unlikely to come into widespread use, he said. It costs about $500,000, and giant aneurysms like Mr. Ratuszny's are rare. Dr. Langer estimated that no more than 1,000 patients a year in the United States would need operations like Mr. Ratuszny's.

The equipment is made by Elana, a company started by the University Medical Center in Utrecht, where Dr. Tulleken teaches. He owns no stock, he said, but relatives do, as does Dr. Langer.

Three million to six million people in the United States have brain aneurysms but do not know it, according to the Brain Aneurysm Foundation in Boston. Aneurysms form when artery walls weaken, but the underlying cause is unknown. Most do not rupture.

But 30,000 people a year do suffer ruptures, with dreadful results. Half die within a month, and many survivors wind up with significant brain damage.

In Mr. Ratuszny's case, the problem seemed to come out of nowhere. He had always been healthy. A soft-spoken, powerfully built man who works out, he has been a lifeguard at ocean beaches and served in the Army Reserves. Now, he works as a Lexus mechanic. He is recently divorced and dotes on his son, Sam, a 3-year-old with a mohawk who shares his father's solid physique and knack for taking things apart.

One morning two years ago, when he was 24, Mr. Ratuszny woke up with an excruciating pain in his head. At first, the diagnosis was migraine, but when the usual drugs did not help, doctors ordered an M.R.I. scan.

By the time Mr. Ratuszny got home from the scanning center, he had five telephone messages waiting—telling him to go straight to the emergency room.

He had what doctors call a giant aneurysm. A three-inch length of an artery had ballooned out to several times its normal diameter and coiled back on itself to form a tangled mass the size of a golf ball inside his head. The vessel was an especially sensitive one: the left internal carotid artery, which feeds the brain centers that control the right hand and create speech and personality.

Mr. Ratuszny was sent to Dr. Langer, the director of cerebrovascular neurosurgery at St. Luke's-Roosevelt, Beth Israel and Long Island College Hospital.

The only way to fix such a large aneurysm would be to bypass it—create a detour for blood to flow around it—by taking a vein from Mr. Ratuszny's leg and sewing its ends to the artery on either side of the aneurysm. Once the bypass was in place, the aneurysm could be sealed off with clips or stitches. It would gradually shrink.

But the operation was risky. The bypass would run from the carotid artery in the neck up over the brain and then down through the Sylvian fissure between the frontal and temporal lobes, to attach to a brain artery beyond the aneurysm. The standard operation would require cutting a hole in the brain artery and then sewing an open end of the bypass vein to the hole—like making a T-shaped junction between pipes.

But to cut an artery, the surgeon must temporarily clamp it, or the patient will bleed to death. The clamps may have to stay on for a half-hour or even an hour. And that is where the risk comes in: cutting off blood flow to the brain can cause a stroke that leaves permanent damage.

Some patients can tolerate the clamping because they have other blood vessels that will fill in for the artery. But Mr. Ratuszny seemed to lack those collateral vessels. Dr. Langer thought he had a high risk of a serious complication like a stroke from the operation—at least 10 percent to 15 percent. And yet the risk of doing nothing was even worse: for giant aneurysms, studies put the odds of rupture or death in one to five years at 50 percent.

Dr. Langer thought Mr. Ratuszny was a perfect candidate for Dr. Tulleken's technique. Not only would it spare him the clamping, but it would allow Dr. Langer to attach the bypass directly to the left internal carotid, which he considered a better repair method than the standard operation. But the laser was not yet available in the United States.

Mr. Ratuszny's aneurysm appeared stable, and Dr. Langer thought it would be safe to postpone the operation until the Food and Drug Administration allowed him to use the laser in a study. Mr. Ratuszny agreed to wait, hoping for a safer operation, even though the aneurysm was causing double vision and tremendous pain in his head that sometimes put him in the hospital.

Dr. Tulleken, gaunt and wry at 66, is a man of formidable eyebrows, and a fan of Spinoza and *The New York Review of Books*. He spends one day a week in the laboratory practicing microsurgical techniques, and he believes that neurosurgery should not be "rude," because the brain does not like being manhandled or having its blood supply clamped off.

This belief led him to devise a new technique. The idea is deceptively simple: instead of cutting a hole in the brain artery and then sewing a vein to it, he sews first and cuts later. That way, the artery does not have to be temporarily clamped, and blood flow to the brain is not cut off. A excimer laser is used to make the hole because it can be slipped into a tight space on the tip of a slender tube and makes a clean cut that stays open without burning nearby tissue.

Late in November, Dr. Langer was shocked to see that Mr. Ratuszny's aneurysm had expanded markedly. It was pressing dangerously on his optic nerve and bulging into his nasal sinus, where it had actually eaten through a bony wall. Mr. Ratuszny's left eyelid drooped, light hurt his eye and he had such severe pain in the eye, face, neck and head that it sometimes made him vomit.

The artery was stretched thin. Dr. Langer ordered Mr. Ratuszny to head for the hospital if his nose began to bleed, because it could be the first sign of a hemorrhage.

The operation could not be postponed any longer. Mr. Ratuszny's father was prepared to take out a second mortgage on his house to pay to have the surgery in Utrecht, but the F.D.A. allowed Dr. Langer to use the device this one time.

A few days before the operation, Mr. Ratuszny said he was eager to get it over with. "If that thing blows up in my head, it's not something I'm going to survive," he said.

Dr. Langer said, "The best case is he goes back to work in about a month and can be a dad, for the rest of his life."

At 2:40 P.M. last Tuesday, everyone in the operating room was ordered to put on safety glasses. A two-minute countdown was begun by Michael Münker, a physicist from Elana, the Dutch company that makes the laser-tipped tubes.

"Thirty seconds left," he called. "Fifteen seconds. Five seconds."

It was not quite *Star Wars.* The laser fired—invisibly. All eyes were on monitors that showed a magnified image of the surgical field. As Dr. Langer withdrew the laser, a flap of tissue cut from the artery wall was stuck to the tube and blood began to flow. The artery was open.

Working through the microscope, using long forceps to grip a fine, curved needle, Dr. Tulleken began the delicate task of sewing the ends of a vein together to complete the bypass. A resident watched, awed by his deft hands.

By 5 P.M., Dr. Michael Tobias, a neurosurgery resident, was fastening metal plates to Mr. Ratuszny's skull with a screwdriver to replace a 4-inch-by-2-inch oval of bone that had been cut out with a saw.

At 6 P.M., the anesthesiologist, Dr. Jonathan Lesser, prepared to wake Mr. Ratuszny, who had been under anesthesia for more than nine hours.

For brain surgeons, the biggest worry comes not during the operation, but after. They watch the waking patient with hope and dread, searching but not wanting to find signs of a stroke. Can he talk? Move his limbs? Respond to commands?

Almost as if he were afraid to watch, Dr. Langer rested on a stool, leaning against the wall, his head bowed. He seemed unaware that he was bouncing his foot in time with a beeping monitor, matching Mr. Ratuszny's every heartbeat.

"This is the painful part," he said. "Sometimes you do everything right in neurosurgery and the patient doesn't do well."

He had predicted that Mr. Ratuszny would most likely have some speech problems after the operation from brain swelling, but that they would be transient.

"Chris!" Dr. Lesser called loudly, standing beside the operating table. "Open your eyes, big guy!"

It took a few more rounds of yelling, but Mr. Ratuszny began to respond. His left knee rose.

"They always move the leg you're not worried about," Dr. Langer said.

But within moments, Mr. Ratuszny was moving all his limbs and even raising his head and shoulders, as if he might bolt up off the table. Dr. Langer leapt from the stool to his side, and he and Dr. Tobias joined the chorus: Squeeze my hand! Stick out your tongue! Groggily, Mr. Ratuszny obeyed. He mumbled a few words in answer to questions, then began shivering violently. The doctors called for extra blankets.

"Chris, you did great," Dr. Langer said. "You're all done, buddy."

As predicted, the day after the operation Mr. Ratuszny did have some speech trouble: he repeated himself and had difficulty finding the right words. But he spoke fluently and laughed at jokes, and the problems began to diminish over the next few days. In his hospital room last Friday, three days after the operation, Mr. Ratuszny greeted visitors cheerfully and said his eye pain had already decreased. By Monday, he was up and about, despite a painful infection in one arm from an intravenous line. He couldn't wait to go home, see his son and return to work.

JEROME GROOPMAN

Being There

FROM *THE NEW YORKER*

*A new movement is arising in American hospitals that advocates let-
ting relatives be present in emergency rooms while doctors try to
resuscitate their loved ones. Pitting chaplains and nurses, who argue
for the emotional needs of the families, against trauma surgeons,
who want to provide appropriate care without potential distractions, the movement is gaining attention. Jerome Groopman, himself
a doctor and medical researcher, takes a first-hand look at how
"family presence" works.*

ONE AFTERNOON IN 1982, a twenty-eight-year-old
Michigan state trooper named Craig Scott stopped a speed-
ing car on U.S. Route 127, outside Jackson. Scott discovered
that the car, a Camaro, had been stolen, and arrested the driver. As
Scott was helping the driver into the back of his patrol car, a passen-

ger in the Camaro pulled out a .38-calibre revolver and shot the trooper three times in the back. Bleeding profusely and gasping for breath, he was taken by ambulance to Foote Hospital, in Jackson. As doctors tended to Scott in a trauma room, his wife arrived at the hospital, and so did several of his colleagues. In the lobby of the emergency room, Scott's wife pleaded with the hospital's chaplain, Reverend Hank Post, to let her see her husband, and Post agreed to convey her request to the physician in charge of the trooper's care.

"We debated back and forth," Post recalled. "The staff was very uncomfortable." Finally, the doctor went to Scott's wife and explained that the hospital prohibited family members from attending resuscitations. But she continued to insist, and eventually the doctor gave in. "It was hard to deny her with all those blue shirts staring you down," said Post, who accompanied the woman to her husband's bedside.

They watched as Scott was given blood transfusions and sent to the operating room for surgery, where he died. Over the next week, Post talked to doctors and nurses at the hospital about allowing patients' family members to accompany them inside the E.R. The doctors and nurses dismissed the idea; many argued that laypeople would have trouble coping with the stress of witnessing resuscitation efforts and patients' deaths. Post disagreed. Ordained in the Christian Reformed Church, a Protestant denomination, he was inspired by the Calvinist tradition of challenging prevailing dogma. He told me that he regarded his effort to open Foote Hospital's emergency room to families as a campaign for "human rights."

For several years, whenever Post was on call, he urged physicians to let patients' relatives sit in on resuscitation attempts, and, when doctors agreed, he stood with the family members next to the patient's bed. "It moved grieving along," Post said. "The families saw quickly how hopeless things were, and, by being present, the family can own part of what went on." In 1985, Post sent surveys to seventy people who had witnessed resuscitation attempts in Foote Hospital's E.R. Among the questions he asked was "Would you choose to participate again if the opportunity were presented to you?" Forty-four of the forty-seven respondents said yes. Several added that although the experience had been unpleasant, it had helped them come to terms with a relative's death.

In 2003, emergency rooms in the United States treated nearly a hundred and fourteen million people; about one in every hundred received CPR or underwent another kind of resuscitation procedure. Resuscitations are gruesome—physicians occasionally have to split ribs or cut into a windpipe in an effort to keep someone who is bleeding or unconscious alive—and just fifteen per cent, at most, are successful. Foote was one of the first American hospitals to permit patients' relatives to witness these procedures, but the practice, which is known as "family presence," is spreading, promoted in many instances by chaplains and nurses over the objections of doctors. (There are no reliable data, but advocates estimate that as many as half of American hospitals allow some form of family presence.)

IN 1993, AT THE ANNUAL MEETING of the Emergency Nurses Association, Patricia Howard, an emergency nurse from Kentucky, submitted a resolution to the group's general assembly endorsing the policy on the ground that "when family members are prohibited from visiting before . . . death, the grief process may be hampered and left unresolved." To her surprise—the association had never discussed family presence—the resolution passed by a large majority. "We've always taken excellent clinical care, but not always excellent psychosocial care," Howard, who just finished a term as the association's president, told me. Family members who witness a resuscitation can help decide when to end efforts to revive the patients, she said. "We have had incidents where families say, 'O.K., you've done enough.' " Like many proponents of family presence, she argues that today Americans are better prepared for the gore of resuscitations than they were ten years ago, because they've seen realistic imitations of such procedures on television. " 'ER' and a lot of graphic programs have made the difference in terms of public expectations and knowledge," she said.

"ER" debuted on NBC on September 19, 1994; two years later, more than thirty million Americans were watching the program each week. Unlike previous medical dramas, such as "Dr. Kildare" and "Ben Casey," from the nineteen-sixties, which featured doctors in pristine white lab coats calmly talking at patients' bedsides, "ER" and its competitors, "Chicago Hope" and "Rescue 911," purported to depict emergency medicine as it is actually practiced. "ER," which was set in a fictitious Chicago teaching hospital, incorporated abundant medical jargon and relied on Steadicams—cameras mounted on the body—and multiple plotlines to create an aura of authenticity. (A typical episode featured patients suffering from severe asthma, premature labor, injuries sustained in a car accident, drug addiction, a genital rash, and heart failure.) In 2005, the American College of Emergency Physicians presented the series with an award for "educating the public about critical issues."

In 1996, researchers at Duke University Medical Center and at the University of Chicago analyzed ninety-seven episodes of "ER," "Chicago Hope," and "Rescue 911" and found that of the sixty patients who underwent resuscitations on the shows nearly two-thirds on "ER" and "Chicago Hope," and all on "Rescue 911," survived. Only one patient on the programs, a sixteen-year-old boy on "Rescue 911" who had inhaled toxic chemicals, endured lasting complications; in fact, patients who survive resuscitations often have brain damage or debilitating neurological conditions. The researchers, who published their findings in an article in *The New England Journal of Medicine*, maintained that "Rescue 911," in particular, tended to feature "miracle cases": younger people surviving acute injuries, rather than elderly patients with chronic heart and lung disease, who account for the majority of resuscitation patients. "The survival rates in our study are significantly higher than the most optimistic survival rates in the medical literature, and the portrayal of CPR on television may lead the viewing public to have an unrealistic impression of CPR and its chances for success," the researchers wrote.

A few years later, two doctors and a professor at the Brody School of Medicine at East Carolina University, in Greenville, North Carolina, sent four hundred surveys to local churches asking members about their views on resuscitation. Nearly half of the two hundred and sixty-nine respondents cited television as a primary source of their information, and predicted, on average, that the survival rate from CPR was seventy per cent. In an analysis published in *Academic Emergency Medicine*, in 2000, the survey's authors cited the 1996 television study, writing that their own research "confirms that the events on these types of shows may be shaping and fueling the public's misconceptions of the effectiveness of CPR."

At the same time, patients and their families have become increasingly involved—and influential—in all aspects of medical care. In the mid-eighties, as the first anti-viral drugs for treating AIDS were being developed, activists demanded to participate in the design of clinical trials directed by the National Institutes of Health and pharmaceutical companies. Inspired by the activists' example, breast-cancer patient-advocacy groups made similar requests. The AIDS groups interrupted meetings and staged "die-ins" at the N.I.H., and, eventually, the physicians in charge of planning the clinical trials agreed to their demands. Laypeople now routinely sit on committees at the N.I.H. and on hospitals' institutional review boards, which assess the ethicality and scientific merit of clinical trials, particularly those involving experimental drugs or procedures. Yet family presence in emergency rooms, which is part of this larger trend, remains controversial. Not only does it represent an incursion by the public into medicine's inner sanctum; more than any other recent development, it reveals the extent to which the power to decide how medicine is practiced is no longer an exclusive prerogative of doctors.

THE FIRST EMERGENCY ROOMS were created about a hundred years ago, when hospitals began designating specially equipped "accident rooms" for the treatment of patients with severe injuries and illnesses, as well as for those who couldn't afford a family doctor.

According to Brian Zink, an associate professor of emergency medicine at the University of Michigan and the author of *Anyone, Anything, Anytime: A History of Emergency Medicine* (2005), surgeons were largely responsible for the design of modern emergency rooms, which typically resembled O.R.s, with bright overhead lights and beds separated by movable partitions. Partly in order to maintain a sterile environment in the rooms, access was strictly controlled. "De facto, family members were excluded," Zink said.

After the Second World War, thousands of families moved from rural areas to cities, where there were not enough general practitioners to treat them, and by the middle of the century hospitals had taken over from family doctors the responsibility of caring not only for the seriously ill and injured but also for the dying. (Until the nineteen-sixties, many ambulances were simply repainted hearses, which were owned and operated by funeral homes.) As Gabe Kelen, the chairman of the Department of Emergency Medicine at Johns Hopkins University, put it, "Death became sanitized, and in the hospital families were spared from seeing the agony of dying." This is precisely what advocates of family presence want to change. (Policies regarding sick or dying children have traditionally been more flexible; many hospitals allow parents to remain with a child during at least some emergency medical procedures.)

Four years ago, Massachusetts General Hospital became one of the first hospitals in Boston to adopt guidelines on family presence. One afternoon recently, Ann Marie, a forty-seven-year-old woman, rode to M.G.H. in an ambulance with her father-in-law, Daniel, who had collapsed in a parking lot behind his apartment building in East Boston. (Names have been changed to protect the family's privacy.) Daniel was seventy-one and had recently spent a month at the hospital, where he had been treated for congestive heart failure. When he collapsed, he had been walking home from a local racetrack, where he had placed some bets. A letter carrier saw him fall and called 911. A team of firemen arrived and administered electric shocks to Daniel's chest in an effort to jump-start his heart. Then medics arrived in an ambulance and continued to try to revive him. A resident of Daniel's building saw the ambulance and called Daniel's ex-wife, who called Ann Marie. By the time Ann Marie got to the parking lot, the medics were preparing to lift Daniel into the ambulance on a stretcher. On the way to M.G.H., the medics gave Daniel oxygen through a tube inserted into his throat.

Patricia Mian, a psychiatric clinical nurse specialist, met the ambulance at the hospital. As the medics wheeled Daniel to a bed in the trauma area, Mian recalled, she told Ann Marie, "If you would like to be with him, you can. I will be with you as well, explaining everything that happens." Ann Marie said that she wanted to stay with Daniel, so Mian escorted her to the head of his bed.

Ann Marie remembered watching as one of the medics told the resuscitation team about Daniel's condition. A nurse named Eric Driscoll took notes on a clipboard, while another nurse cut away Daniel's clothes with a pair of shears. Keith Marill, an attending emergency-department physician, stood at Daniel's right hip; Kriti Bhatia, the senior resident in charge of the resuscitation, stood at the foot of the bed. A medical intern and a surgical resident were also present.

Daniel's skin was blue from the chest up, indicating that his circulation was impaired, and his pupils were dilated.

"We need central access," Bhatia remembered instructing the surgeon. "Please put in a femoral line." The surgeon splashed an iodine solution over Daniel's groin and pressed a gloved finger into the cleft between his lower abdomen and his left thigh, searching for a pulse in the femoral artery.

"He's pulseless," Marill recalled saying, palpating the same area on Daniel's right side. "Try your best to hit the vein."

The surgeon stuck a large-bore needle into Daniel's groin several times before he found the femoral vein. Then he threaded a long catheter through the vein into the inferior vena cava, a large vessel that rises from the abdomen to the heart. A nurse attached a bag of saline solution to the catheter.

As the intern compressed Daniel's chest, the nurse forced oxygen into his lungs by squeezing a balloon-shaped bag that was attached to the tube in his trachea. Gradually, Daniel's blue skin turned ivory. Ann Marie placed her hands on his head to comfort him. She noticed that his eyes were wide open and unblinking, which upset her. She glanced at Bhatia but, fearful of distracting her, didn't say anything.

Mian encouraged Ann Marie to talk to Daniel. "It's O.K., Dad," Ann Marie told him. "It's O.K. I'm here."

"He's in P.E.A.," Bhatia said, looking at a monitor by the bed. "Give him epinephrine and atropine." P.E.A. stands for pulseless electrical activity; Daniel's heart muscle was unable to pump enough blood to generate a pulse.

Ann Marie saw a nurse inject the medicines, but she didn't catch the names of the drugs or understand what they were for: to help Daniel's heart contract effectively and move blood faster through his vessels. Instead, she focussed on Daniel's face. Each time the intern pressed on Daniel's chest, it seemed to her that his bulging eyes might jump out of his head.

"His potassium may be high," Bhatia said. "I want a bolus of D50, ten units of insulin, with calcium and bicarb." (D50 is a highly concentrated glucose solution that, in the presence of insulin, will help draw excess potassium from the blood into the cells; calcium helps the heart contract.)

"He has a pulse!" Marill announced. The intern stopped pumping on Daniel's chest.

"His EKG shows a heart rate of eighty-four," Bhatia told the team. Driscoll noted that the resuscitation attempt had been under way for five minutes when Daniel achieved a stable blood pressure of a hundred and twenty-five over seventy-two.

Ann Marie had no idea that Daniel had revived. "I didn't know that his heartbeat had ever returned," she told me later.

Daniel sustained the blood pressure for seven minutes. Then Bhatia said, "Oh God, he lost his pulse."

"He's got a wide complex with an accelerated ventricular escape rhythm," Marill said, looking at the monitor. Once again, Daniel's heart had ceased to beat effectively.

A nurse administered more epinephrine and atropine, and Driscoll made notes on his clipboard.

"He hasn't got his pulse back," Marill said.

The intern began to sweat as he pressed on Daniel's chest.

"Keep pumping," Bhatia said. "We need an ultrasound." She lowered a wand over Daniel's chest which bounced sound waves off his heart. One cause of pulseless electrical activity is cardiac tamponade, a condition in which fluid accumulates around the heart and compresses it like a vise, preventing blood from entering the organ. The ultrasound would detect tamponade and show the strength of any muscular contractions.

"Does he have any cardiac activity?" Bhatia asked.

"Nothing happening," Marill said as he examined the ultrasound image. Daniel's heart muscle was flaccid.

"Does anyone have any other thoughts, suggestions, or is there anything else we should do?" Bhatia asked.

The doctors and nurses looked at her in silence.

"O.K., I guess this is it, then," Bhatia said. Marill placed his stethoscope over Daniel's chest and listened for breath sounds. There were none. He felt Daniel's neck for a pulse and shook his head. Finally, he shined a penlight in Daniel's eyes. The pupils didn't respond.

Driscoll noted that Daniel was declared dead at 2:35 P.M., seventeen minutes after he arrived at the E.R.

Later, Ann Marie said that she thought that Daniel had been dead for a long time before the doctors and nurses stopped their efforts. She was troubled by his fixed stare, which made her feel that Daniel wasn't at peace, and, at her request, Mian asked a nurse to close his lids.

Ann Marie told me that Daniel "was the third family member to die in my arms." In 1974, she had been at her father's bedside at M.G.H. when he died, from complications of advanced diabetes, and she had cared for her mother until her death, ten months later, from brain cancer. Ann Marie said that she wasn't sure why she had decided to witness her father-in-law's resuscitation. "I think he was probably already dead in the parking lot," she said. "But it's so sad to die alone. I wouldn't want to die alone. I didn't want him to die alone. And when I was in the room I knew the doctors and nurses did everything they could."

I asked Ann Marie how she felt about what she had seen in the emergency room. "I used to be very sensitive," she said. "My mother's and father's deaths made me stronger." Nevertheless, she added that she often pictured Daniel's eyes "jumping" in synch with the intern's chest compressions. "The whole thing was traumatic for me," Ann Marie said. "But I try to bypass it by making myself think of good things."

FEW ATTEMPTS HAVE BEEN MADE to measure the psychological impact of family presence, either on patients' families or on doctors and nurses in the E.R. Most of the existing studies consist of surveys and involve so few people that they cannot be considered significant. Addenbrooke's Hospital, in Cambridge, England, instituted a policy of family presence after completing a single study, in 1998, involving twenty-five relatives of people who had undergone a resuscitation at the hospital. Thirteen relatives were invited to witness the procedure (eleven chose to do so); twelve, who formed the control group, were not. Researchers surveyed eight relatives who had watched a family member die, at three months after the procedure and again three months later, and found no evidence of trauma. Moreover, these relatives said that they were pleased with their decision to observe the resuscitation. Despite the study's small size, it has been cited repeatedly in medical and nursing journals as proof of the therapeutic value of family presence.

In August, 2000, the journal *Circulation* published new guidelines for emergency cardiovascular care from the American Heart Association, which recommended that family members be allowed to witness resuscitation attempts. Two months later, at the annual meeting of the American College of Chest Physicians, in San Francisco, researchers from Tripler Army Medical Center, in Honolulu, distributed a survey about family presence to the attendees, who included physicians, nurses, and health-care workers such as respiratory therapists. Of five hundred and fifty-four respondents, the majority (seventy-eight per cent) opposed family-witnessed CPR for adults. More physicians (eighty per cent) than nurses (fifty-seven per cent) disapproved of the practice. The researchers, who published an analysis of the survey in *Chest*, a leading journal of thoracic medicine, found that where the respondents lived was a better predictor of their attitude toward family presence than the size or type of hospital in which they worked. Health-care workers in the Midwest were most likely to favor the practice (thirty-seven per cent); those in the Northeast least likely (twelve per cent). "While the reasons for these differences are unproven, we speculate that ten years of efforts by the Foote Hospital (Jackson, MI) staff may have taken root in the Midwest, making FWR"—family-witnessed resuscitation—"more acceptable in that region," the researchers wrote.

One group whose members have actively opposed family presence is the American Association for the Surgery of Trauma. In 1999, a team of medical researchers led by R. Stephen Smith, a trauma surgeon and a professor at the University of Kansas School of Medicine, in Wichita, surveyed three hundred and sixty-eight trauma surgeons and twelve hundred and sixty-one emergency nurses to learn their views of the practice. The respondents were asked to agree or disagree with statements such as "I would want to be present during TR"—trauma resuscitation—"following injury of a member of my family." Almost all of the nurses said they agreed with the statement; almost all of the trauma surgeons said they did not. Sixty per cent of the nurses approved of family presence during cardiopulmonary resuscitation and ninety-seven per cent of the surgeons disapproved, saying that it interfered with patient care and increased stress on the doctor.

In a 1999 report for the A.A.S.T. summarizing the results of the survey, Smith and four co-authors cited Federal Aviation Administration regulations instituted in 1981 to reduce the number of airplane accidents caused by distracted pilots. Known informally as the "sterile cockpit rules," the regulations prohibit crew members from engaging in activity that is not "required for the safe operation of the aircraft"—and which the F.A.A. defined as unnecessary conversation, eating, reading, radio communications, and public-address announcements. Smith and his co-authors argued that resuscitations involve tasks as demanding as those required to fly a plane. Like pilots, they wrote, emergency-room teams must assimilate large quantities of data in a short time and make quick decisions; potential distractions, such as the presence of a family member, could jeopardize the success of a resuscitation.

Smith argues that the debate over family presence has exposed a conflict in medicine between, on one side, chaplains and nurses, who worry about families' emotional needs, and, on the other, physicians, who are primarily concerned about the quality of clinical care. In this sense, the debate is a sign of how power has shifted within the hospital; a movement led by chaplains and nurses to change a long-standing medical protocol would have been inconceivable when I was a resident in the emergency room thirty years ago. (Chaplains' tasks were limited to assisting patients with prayer and delivering last rites, and nurses, particularly in surgery, were viewed as handmaidens who should take orders but never give them.)

Smith's hospital has addressed the conflict by devising a compromise: patients' families are allowed to join them in the E.R., but only after invasive procedures have been completed and at the discretion of the trauma surgeon. Even so, Smith said, a family member's presence has sometimes been disruptive. In one case, a woman suffering from multiple fractures and internal injuries was brought to the hospital by her husband, who wanted to join her in the E.R. Smith and the other physicians who were treating the woman suspected that she might be a victim of domestic abuse and worried that she would be reluctant to recount her medical history in front of her husband.

Eventually, the man was persuaded to remain in a waiting room. (The doctors determined that the woman had indeed been abused.)

Smith believes that hospitals should retain the right to invite a patient's relatives into the E.R. on a case-by-case basis. At the same time, he said, laypeople need to realize that they may not understand much of what they see there. "Even medical students don't know why certain things are done during trauma resuscitation," he said. "It's like me taking a tour of a nuclear power plant."

THE HOSPITAL WHERE I PRACTICE, Beth Israel Deaconess Medical Center, in Boston, does not encourage patients' family members to witness resuscitations, though there is no official policy. Lachlan Forrow, an internist who directs the hospital's ethics programs, said that some doctors have favored family presence on the theory that if a patient's relatives saw how violent resuscitations can be they would be more inclined to agree to end efforts to keep the patient alive. But, he said, there is no solid evidence to support this view. A more fundamental problem, Forrow argues, is that laypeople don't necessarily trust doctors to make the right decision. "If the symptom of lack of trust is addressed by having blood splattering on the walls, then as physicians we've gotten ourselves into some kind of weird dynamic," he said. Forrow points out that there is no proof that witnessing a failed resuscitation is more therapeutic for a grieving family member than being told about it afterward by a physician. "Seeing a health-care provider inflict physical pain or disfigure the body of a loved one?" he said. "It seems to me that we should be able to accomplish transparency without that degree of brutality."

Keeping families out of the emergency room, however, ultimately may be impossible. "We are entering an era of openness in every field," Alasdair Conn, the chief of emergency services at M.G.H., told me. "You want to know whether your stockbroker is a good broker. You want to get a second opinion on a legal decision. It's happening in medicine, too. In many medical situations, there is no one right way to do things. There is this questioning, a search for alternative answers. It's different now, coming out in the waiting room and saying, 'We did everything possible, and your father died.' Then the families ask, 'Did you give the drugs? Did you do this? Did you do that?'"

Conn said that he had opposed family presence at M.G.H. until he imagined himself as a patient's distraught relative. "If my daughter or my wife or any of my relatives were in pain and in the emergency department, I would want to be there with them," he told me. "Even when they start putting in lines, or taking blood, or putting in a chest tube, or doing a resuscitation.

"Let me distill it," Conn continued. "Suppose you had the opportunity to spend the last three minutes on earth with your wife. She was just brought into the emergency room. She is semi-conscious, and she is probably going to die. Would you want to say a few words to her, or would you rather be someplace else? I think most people would say that they want to be with her."

MATTHEW CHAPMAN

God or Gorilla

FROM *HARPER'S*

In 2005, eighty years after the Scopes Trial, a courtroom in Pennsylvania became the scene of another famous legal fight over evolution, this one involving the Dover School Board and the proposed teaching of "intelligent design." Matthew Chapman, a descendant of none other than Charles Darwin himself, was there to watch how his forebear's theory fared in the American justice system.

I N THE CASE of *Kitzmiller v. Dover Area School District*, eleven parents sued to remove intelligent design from the curriculum. The defendants brought in some of the leading lights in the intelligent design movement to defend it as science and elucidate the gaps in evolution. The plaintiffs brought in experts on evolution to explain it and refute intelligent design. That's the basic story, but if you think you know everything there is to know about this, you are wrong. Only I know the truth.

Dover lies a mere thirty miles from the Three Mile Island nuclear plant, and the meltdown of its core and subsequent leak in the Seventies is responsible for the weird behavior now seen in the locals.

I have no evidence for this belief, and my lack of evidence is a matter of pride.

Having said that, I suppose I should declare my bias at the start. My great-great-grandfather was Charles Darwin. This was not something I thought much about growing up in England. Evolution was fully accepted. Darwin was a historical figure. If I did think about my connection to him, it was only negatively. The pressure to succeed academically and the unlikeliness of doing so in comparison to my ancestor was such that I decided to turn my back on academia and pursue a course of willful ignorance. When I finally moved to Hollywood in the early Eighties, I had gone about as far as I could in that direction.

I then discovered that many Americans not only rejected the theory of evolution; they reviled it. I had come here in part because I never felt comfortable in England. I hated the snobbery and thought of America as being less weighed down by its past, more advanced. Sir Francis Drake might have been the first man to sail around the world, but it was an American who first set foot on the moon. Now here I was in the New World faced with a willful ignorance that went far beyond anything I had ever attempted.

True, I did not know much about evolution, but a quick study of the subject showed that 99 percent of scientists believed in it. Why would one doubt them? Did the pedestrian question the theory of gravity? Did the farmer who went to the doctor question his diagnosis? Why in this one area of science did nonexperts feel compelled to disagree with those who clearly knew better?

Dover's population, with an influx of people who commute to nearby towns, is approaching 2,000. The Dover Area School District, however, covers a largely rural population of about 24,000, and Dover Senior High School has about 1,000 students.

In June of 2004, reporters Joe Maldonado and Heidi Bernhard-

Bubb, working respectively for the *York Daily Record* and the *York Dispatch*, covered a school board meeting in Dover. Under consideration was a new edition of *Biology: The Living Science*, by Kenneth Miller and Joseph Levine. The chair of the curriculum committee was Bill Buckingham, an ex-cop and corrections officer and self-confessed OxyContin addict. According to Joe and Heidi, he told the meeting that he was disinclined to purchase the book because it was "laced with Darwinism." He went on to say, again according to the reporters, that "it's inexcusable to teach from a book that says man descended from apes and monkeys." The separation of church and state, he continued, was "mythical," and he wanted a book that included views of creationism as well as evolution. When asked after the meeting what consideration he intended to give to other religions, he said, "This country wasn't founded on Muslim beliefs or evolution. This country was founded on Christianity, and our students should be taught as such."

The following Monday, at another meeting, Buckingham apologized for his comments but went on to grumble that "liberals in black robes" were taking away the rights of Christians. Bill, who, from the record, seemed to be alternately menacing and self-pityingly apologetic, finally cried out, "Two thousand years ago someone died on a cross! Can't someone take a stand for him?" Fellow creationist and school board president Alan Bonsell, owner of a nearby radiator and auto-repair shop, supported Buckingham's ideas in a more reasonable tone, and conflict ensued. There were accusations of atheism and un-Americanism, and many tears were shed.

But Buckingham and Bonsell were undeterred and soon fixed on the intelligent design screed *Of Pandas and People* as the book they wanted the ninth-grade students to have in order to get some "balance" in their science education. There were votes and more votes (and more tears), and finally *Pandas* was voted out. But someone still wanted the book to be available to the students, and an anonymous donation of sixty books was made to the Dover High library.

It was eventually agreed that a statement would be read to the ninth-grade science students before they began studying evolution that read in part:

> Because Darwin's Theory is a theory, it continues to be tested as new evidence is discovered. The Theory is not a fact. Gaps in the Theory exist for which there is no evidence. A theory is defined as a well-tested explanation that unifies a broad range of observations.
>
> Intelligent Design is an explanation of the origin of life that differs from Darwin's view. The reference book, *Of Pandas and People*, is available for students who might be interested in gaining an understanding of what Intelligent Design actually involves.

The science teachers refused to read this, so Superintendent Richard Nilsen and Assistant Superintendent Mike Baksa went from classroom to classroom and made sure every ninth grader got to hear it.

On December 14, 2004, eleven Dover parents, represented by the Pennsylvania ACLU, Americans United for Separation of Church and State, and the powerful Philadelphia law firm Pepper Hamilton, filed suit in Federal District Court in Harrisburg.

The Comfort Inn, where I stayed during the six weeks of the trial, is in downtown Harrisburg. It overlooks the Susquehanna River and a series of beautiful bridges that cross it. A cooling breeze blows off the river but never enters the hotel. The windows are sealed shut. Your climatic choices are limited to Off, Fan, Low Heat, High Heat, Low Cool, and High Cool. This became, to my mind, a perfect metaphor for the debate.

The case was a civil suit without a jury, so members of the press were given the jury box to sit in. Placed on one side of the modern courtroom, these were the best seats in the house, comfortable leather chairs affording great views of a screen upon which exhibits would be displayed. To our left was the witness box, and beyond it, the bench occupied by Judge John E. Jones III, a good-looking man of fifty. In front of him sat the clerk of the court and the stenographer. Right in front of us was the lectern from which the lawyers asked their questions. To our right were the spectators in the back of the court on two rows of uncomfortable wooden pews.

The plaintiffs made their case first. Seeking to keep the judge—a Bush appointee—engaged, both sides cut back and forth between the loftier theses and the human beings who drove them.

During the first two days of the trial, for example, Ken Miller, a professor of biology at Brown University, co-author of the biology textbook now used at Dover High, and an expert on "the coupling factor on the thylakoid membrane," was followed by office manager Tammy Kitzmiller, a pretty, divorced mother of two, whose name was attached to the suit because one of her daughters was actually in the ninth grade.

I eventually got to meet Tammy and her teenage daughters. The daughters had numerous piercings in their ears. Tammy had a belly ring. I did not interview Ken Miller, but I suspect he does not have any piercings; however, if you read his testimony (available on the National Center for Science Education website), you'll get a pretty good overview of the nature and function of science. Like many of the plaintiffs' witnesses, Miller, a practicing Catholic, had no trouble believing in both evolution and God.

On the third day of testimony, Robert Pennock took the stand and was questioned by Eric Rothschild, the lead attorney for the plaintiffs. Rothschild is a man in his late thirties with a balding head shaved close. He has a deceptively cherubic face; but it's a dark face too, with the air of someone keeping a secret. One might imagine that as a geeky child he had encountered some bullying and was not about to let it continue into adulthood.

Pennock, an enthusiastic man with a beard, is a professor at Michigan State University. He has a B.A. in biology and philosophy and a Ph.D. in the history and philosophy of science. His primary appointment is in the College of Natural Sciences, but he's also in the Department of Philosophy, the College of Engineering, the Computer Science and Engineering Department, and the Graduate Program in Ecology, Evolutionary Biology, and Behavior.

He spoke of how evolution is "a great exemplar of the scientific method. It's a well-confirmed interlinked series of hypotheses," and is useful not just in and of itself but as a way of learning how to think. "One needs to know it with regard to medicine, and even with regard to engineering applications. . . . So there's practical applications to evolution right now. You can get a job at Google if you know something about evolution."

We next received a lesson on the history of methodological naturalism, going back as far as Hippocrates, who refused to see epilepsy, then known as "the sacred disease," as divine possession but instead looked for natural causes.

This was followed by a critique of intelligent design with particular attention to William Dembski, a big cog in the movement. Pennock read from an article of Dembski's entitled "What Every Theologian Should Know About Creation, Evolution, and Design":

> The view that science must be restricted solely to purposeless naturalistic material processes also has a name. It's called methodological naturalism. So long as methodological naturalism sets the ground rules for how the game of science is played, IDT has no chance.

And later:

> In the words of Vladimir Lenin, "What is to be done?" Design theorists aren't at all bashful about answering this question. The ground rules of science have to be changed.

Rothschild paused a moment and then said, "And I have to admit I didn't know until I read that that Vladimir Lenin was part of the intelligent design movement, but putting that aside . . . "

Soon after this, he received an Internet proposal of marriage.

Pennock's cross-examination was by a man named Patrick Gillen, of the Thomas More Law Center, which had offered its services *pro bono* for the defense. This seemed like a logical inconsistency from the start. Run by Richard Thompson, a Catholic and former Michigan prosecuting attorney who made a name for himself by trying to put Jack Kevorkian in prison, its stated mission is "Defending the Religious Freedom of Christians," "Restoring Time-honored Family Values," and "Protecting the Sanctity of Life," which, as a biblical literalist myself, I take to mean defending such freedoms as the biblically mandated right to capture women in battle, shave their heads, lock them up for a month, rape them into matrimony (Deuteronomy 21:10), and then deny them the right to an abortion afterward.

All well and good, but if the defense thesis was that intelligent design was merely another scientific theory, what were these Catholic activists doing in court?

One of my chief defects is an inability to hate people I violently disagree with once I get to know them. In Gillen's case, my sympathy was ignited by the contrasts in his face. A tallish man in his midthirties, with a long head topped with thinning hair, he had excellent teeth, revealed frequently in a blazing grin; but from the middle of his nose up, he wore an expression of extreme anxiety, his brows furrowed, his eyes filled with concern.

BEFORE GETTING INTO GILLEN'S CROSS of Pennock, I should paint a brief portrait of the two legal teams.

On the plaintiffs' side, apart from Rothschild, a lawyer who up until now had spent most of his life in the corporate environment of reinsurance law, was another lawyer from Pepper Hamilton, Stephen Harvey. The third lawyer, Witold "Vic" Walczak, was from the ACLU. Now and then another lawyer from Pepper Hamilton, Thomas Schmidt, was used to cross-examine defense witnesses who were so clearly feebleminded or old that the sharp-elbowed style of the other three might actually render them unconscious.

Lending intellectual heft to this legal phalanx were Katskee and Matzke. Richard Katskee, a lawyer from Americans United for Separation of Church and State, was an expert in constitutional law. Nick Matzke, from the National Center for Science Education, provided the science.

But the team did not end here. There were two legal assistants and the unsung hero of the plaintiffs' case, Matthew McElvenny, Technology Specialist, the faultless Wizard of Oz whose computer held all the necessary exhibits—drawings of bacteria, excerpts from books and articles, depositions, even news video—and projected them up on the screen.

As anyone will tell you who has covered a trial, sleep is the slyest and most persistent enemy, but when the Wizard of Oz was on, highlighting and scrolling without a single mistake, one inevitably perked up.

Here then was a team of highly skilled professionals operating in an atmosphere of frictionless amiability. Here was a collegiate machine.

On the defense side, one was reminded more of a dysfunctional family with a frequently absent father.

Richard Thompson, who, in profile at least, bore an uncanny resemblance to William Jennings Bryan, was the star, and it was hard to imagine that any case in the history of his Thomas More Law Center had ever been as important as this one. For the first few days, he attended court dutifully, once or twice cross-examining a witness in an odd combative style, often turning toward the jury box (filled with an unsympathetic jury of reporters), then turning back to point his finger at a witness to ask a question whose substance seemed to bear no relationship to the tone in which it was asked. Then he would sit down and rock back and forth in his chair, staring up at the ceiling as if contemplating weightier matters—and then he'd disappear for a week.

Next among the defendants' lawyers—though some say first among them—was Robert Muise. He is a tall, sturdy man, quietly resolute, with a faint Boston accent. Always willing to talk, as unfailingly polite as Gillen and Thompson, he seemed to be a tough guy underneath but worn down, becoming a victim. Perhaps this had nothing to do with politics and religion: he and Gillen, though both only in their late thirties or early forties, had seventeen kids between them, one nine, the other eight. Thompson, perhaps too busy pursuing Doctor Death, had produced a scant three.

For a while there was a legal assistant, but she went the way of Thompson. Sometimes it was nine against one, Gillen alone, smiling dutifully, fumbling for his own documents. By prior arrangement or out of simple human (humanistic?) decency, the plaintiffs' machinery was put at his disposal so that he could display his documents on screen.

GILLEN BEGAN by asking Pennock questions designed to show that just because a theory (such as the Big Bang) confirms some people in their religious beliefs, it is not necessarily unscientific. Pennock quickly sliced this up into its constituent parts and disposed of it. People could believe what they wanted; that was neither his business nor particularly interesting: all that counted was the evidence.

Gillen now moved in on the Ancestor, a computer program that Pennock and three colleagues had designed to demonstrate natural selection. Self-replicating computer organisms are dropped into an artificial digital "life system." The "viruses," if you like, are then seen to mutate and develop, those that adapt best surviving, those that don't dying.

"They evolve things," said Pennock, waving his hands around, "where the programmer would think, 'Why, I would never have even thought to do it that way!'"

Gillen began to ask another question, but Pennock, leaning even farther forward in his chair, now bouncing with enthusiasm, was too full of gusto to be stopped. "And the other thing about it is—sorry, I get excited about this . . . we can keep track of the full evolutionary history! So we have a complete fossil record, if you will!" He beamed at the courtroom, which responded with supportive laughter.

Gillen collected himself and pushed on, trying to extract the obvious: all this might be true, but if anyone looked at one of the resulting "organisms," he would actually be correct if he inferred that there was an intelligent designer behind it—four of them, in fact. Pennock would have none of it. Neither he nor Darwin was interested in who created the original organism (this, of course, was a tough concept for Gillen, who clearly had a pretty good idea who He was and had to bite his tongue not to mention Him by name), only in the mechanism of its development.

When court finished for the day, I asked Pennock if I could come and see these organisms, hoping that there would be some Pac-Man-like creatures to view, but was disappointed to be told that they do not exist in visible form.

I had only one problem with Pennock—and in fact with all the scientists who spoke: their use of unnecessarily obscure words. As if the science wasn't hard enough to follow, Pennock would use a word like "*qua*" instead of "as" or "by virtue of being." For example, he said, "Sometimes people will speak *qua* scientist, and sometimes they will speak about something from their own personal views." I found myself wondering if he talked to his wife like that: "Listen, honey, this place is a mess, and I'm not just saying that *qua* husband."

ONE NIGHT DURING THE TRIAL, a local preacher named Reverend Groves put on a show at the Dover firehouse that consisted of him showing a DVD entitled "More Reasons Evolution Is Stupid." The producer and star of the DVD is a man named Kent Hovind, an ex-science teacher, a.k.a. Doctor Dino, who owns a creationist theme park down in Pensacola, Florida. Hovind would throw up an aspect of evolution (that apes and man share a common ancestor, say), with the addition of enough complex-sounding science to make himself seem well-informed, and then dismiss it with the line "That's stupid!" or "I'm sorry, boys and girls, but that's not common sense, that's just stupid!"

When this endless clay-pigeon shoot was done, and the DVD turned off, a man named Burt Humburg, a medical resident at Penn State, calmly raised up a table-top document stand and started to defend evolution. Within moments, a woman, suffering from dental defects that would do an Appalachian proud, was standing in the middle of the hall shouting, "You've been brainwashed in college!" There were grunts and murmurs of agreement, and Burt, although he struggled on manfully for a while, was silenced. I would catch up with him later and find, increasing my admiration, that he was raised in some charismatic division of the church where they spoke in tongues but had been washed clean by the H_2O of science and born again in reason.

A few days later I interviewed Reverend Groves for a documentary film I was shooting. A wiry little homohater in his late fifties, dressed in tightish pants and cowboy boots, he had an insinuating manner that belied his courage. Every Halloween he joins the parade in York, putting on one of those gruesome anti-abortion shows so beloved of the breed, smashing blood-filled dolls and displaying graphic photographs of aborted fetuses and so scaring the children that in 2002 he was actually arrested by the local police. He sued, however, on the basis of free speech, won, and is now a parade fixture, albeit at the rear.

By this time, it was public knowledge that I was an offspring of Darwin, and in the course of the interview it became apparent to me, really for the first time, how hated the poor old codger is. People such as Groves believe that Darwin marks a point in history from which materialism sprang, bringing with it Hitler, Stalin, Pol Pot, pot, sex, prostitution, abortion, homosexuality, and everything else nasty in the world.

"The moral condition of America," said Groves, "is a result of taking steps away from the Bible and away from God over the past fifty to one hundred years, since evolution was introduced . . . you cast yourself on the sea of nothingness as far as the moral code goes. And every man does that which is right in his own eyes, as the Bible said.

And when you do that, you—it's like a moral free-for-all. And that's what's happened in America. We no longer have religious freedom in America, we have a religious free-for-all in America. America was not that way, was not that immoral when it stayed to its Christian roots originally. . . . And now we're in the purge, with the ACLU, with legal organizations such as that, to purge our whole society from anything Christian."

It occurred to me how lucky we are that Darwin lived such a dull monogamous life. Had he been an adulterer, his theory would be dead and buried. Or maybe not. Joseph Smith, a contemporary of Darwin's and the polygamous founder of the Mormons, simply stated that his "truth" was handed to him on a set of golden plates that then mysteriously disappeared. Perhaps if Darwin had done the same he'd have avoided all this controversy.

According to a recent U.S. poll, 54 percent of American adults now dispute that man developed from earlier species, which is a 10 percent increase since the last poll, in 1994. Scientists must bear some responsibility for this: they just don't seem able to provide entertainment the way the other side can. When did you last hear a scientist come up with anything as fun or contentious as man of God Pat Robertson calling for the assassination of Hugo Chavez? Why haven't we seen a man of science on TV asking Bush to explain why God, being such a great pal, gave him such lousy intelligence on the WMDs, or demanding an explanation for all the gaps and contradictions in the biblical record?

As Groves had shown no restraint in taking a whack at my ancestor, I felt no compunction in whacking back and asked him some of the questions Darrow asked Bryan at the end of the Scopes trial. Was Jonah really swallowed by a whale? Yes. How did Joshua "command the sun to stand still" when we know that the earth goes around the sun and that stopping it would be disastrous? That's what a miracle is. Were the six days of creation literal days, and how old is the earth? Bryan, when pushed, conceded that perhaps the six days could have been symbolic, and on the subject of the age of the earth pleaded a pathetic ignorance. "I have been so well satisfied with the Christian religion," Bryan said, "that I have spent no time trying to find argument against it." But Groves was made of sterner stuff. He was unashamed of a literal reading of Genesis and an earth that was only 6,000 to 10,000 years old. Carbon dating was nonsense. And that was that.

When I visited Groves in his cinder-block church, he had set up his own video camera to film me filming him. He told me it was just to keep a record of the event, and I did not object. At the end of my interview, he asked me if I was an atheist, and I replied that, no, I was an agnostic, believing that faith even in nothing was too much faith. I finished by observing how odd it was that a country as riddled with Christian faith as America has so little regard for its poor, sick, and imprisoned.

Two days later, two reporters told me they had visited the church in search of local color and found me booming from a TV on the altar, declaring my agnosticism to many gasps of horror. Apparently, the consensus was that I'd end up in hell, probably to find Great-Great-Grandpa sitting at the Devil's side.

When I upbraided Groves about this—he had not told me I was to be used in this way—he shrugged off my objections and told me it had been "educational." He and his flock concluded that I had a different understanding of Christianity. Coming from Europe, mine was "more socialistic," while his was more concerned with "individual salvation."

THE FIRST DEFENSE WITNESS, Michael Behe (father of nine), looks like the archetypal professor, bearded, vague, tweed-jacketed. Author of *Darwin's Black Box*, he is a biochemist and professor of biology at Lehigh University in Bethlehem, Pennsylvania. Bethlehem, it turns out, is the birthplace of the expression "irreducible complexity." Bethlehem!

Behe's shtick, if I may so characterize it, is largely to do with the irreducible complexity of the bacterial flagellum. It is slightly more than this, but if you can understand the flagellum argument, you can understand it all.

Although the concept of irreducible complexity is sold as "brand new," it is in fact more "like new." It began with religious philosopher William Paley's 1802 argument about someone finding a watch and inferring that there had to be a watchmaker. The argument now also includes reaching the same conclusion while looking at Mount Rushmore or seeing "John Loves Mary" written in the sand.

The bacterial flagellum is, however, an amazing thing. Without a diagram, it's more or less impossible to describe. Behe had one, at which he pointed with a laser pointer. In fact, he pointed at everything with a laser pointer. Even when there was only text on the screen, often stuff he had written himself, a red dot danced distractingly across the words. Here he is describing the flagellum:

> The bacterial flagellum is quite literally an outboard motor that bacteria use to swim. . . . This part here, which is labeled the filament, is actually the propeller of the bacterial flagellum. The motor is actually a rotary motor. . . . It spins the propeller, which pushes against the liquid in which the bacterium finds itself and, therefore, pushes the bacterium forward through the liquid.
>
> The propeller is attached to something called the drive shaft by another part which is called the hook region which acts as a universal joint. . . . The drive shaft is attached to the motor itself which uses a flow of acid from the outside of the cell to the inside of the cell to power the turning of the motor, much like, say, water flowing over a dam can turn a turbine. . . . It's really much more complex than this. But I think this illustration gets across the point of the purposeful arrangement of parts. Most people who see this and have the function explained to them quickly realize that these parts are ordered for a purpose and, therefore, bespeak design.

I often encountered Behe outside the courtroom. He was a likable man, and when he found out I was a Darwin descendant he was delighted, stating later in a newspaper article that I was a friendly fellow and my presence in the courtroom was a comfort to him. But I could not get past two thoughts. If an intelligent designer had made the bacterial flagellum, it was logical to assume he had made everything else, and if he had, wasn't this by definition God? One day, I was having this debate with him when another man weighed in, suggesting that since complex machines like the space shuttle are designed by a team, wasn't it probable that the flagellum was also made by a team?

Behe smiled tolerantly and shrugged: he himself believed in a single designer, that was his personal opinion; we could believe what we wanted.

My second thought was that if you looked back at the history of science, you could point to any number of things that, given our knowledge at the time, seemed possible only through the intervention of God but that later turned out to have natural explanations even Behe accepted. I missed the point, he told me—and told Rothschild later during cross: the bacterial flagellum is not only complex, it is irreducibly complex. In other words, if you removed one element of it, none of the others had function, and so the whole could not have developed by natural selection but must have been abruptly created with all its parts in place. In this context, the mousetrap was often cited.

On the stand, Behe sat forward in his chair, earnest and concentrated. Only once did I see him lose his composure. This was when Rothschild revealed that Behe's own department at Lehigh had issued a statement saying it fully supported evolutionary theory and that

The sole dissenter from this position, Professor Michael Behe, is a well-known proponent of intelligent design. While we respect Professor Behe's right to express his views, they are his alone and are in no way endorsed by the department. It is our collective position that intelligent design has no basis in science, has not been tested experimentally, and should not be regarded as scientific.

Behe put his hands behind his head and leaned back in his chair, smiling defiantly. He looked like a naughty child who had told his mother he'd seen a ghost and wouldn't budge from the story no matter what. I couldn't help wondering what Behe would be without intelligent design. The scientific community may despise him, but he is beloved on the other side. He gets invited to talk all over the country, and he has sold a lot of books.

OUTSIDERS SUCH AS MYSELF were in a froth of anticipation for the testimony of the pugnacious, OxyContin-addicted crusader Bill Buckingham.

By this time many of the plaintiffs had taken the stand and confirmed press reports of Buckingham's outlandish statements. They had been a diverse group, funny, angry, simple, complicated, intelligent, rich, poor, some eloquent on the Constitution, all but a few of them believers, but all having a clear respect for learning and fairness. A picture had slowly come into focus of an arrogant, brutish fundamentalist who would hold to his beliefs no matter what the consequences.

But when he arrived, walking with a cane, he seemed old, tired, and subdued. If, as Samuel Johnson said, "Patriotism is the last refuge of a scoundrel," Buckingham was upping the ante with his lapel pin, an American flag wrapped around a cross. He had been through two stints of rehab to kick his addiction, and one wondered if another drug had been prescribed to keep him from making outrageous statements in court.

Knowing that Stephen Harvey was about to question him, one almost felt pity. Harvey, a prematurely gray-haired man in possession of the best suits at the trial—and a Republican, it would turn out— was a man whose considerable personal charm and boyish smile disappeared entirely during cross-examination and was replaced by a cold intensity that was almost frightening to behold.

Buckingham, a 1973 graduate of the Penn State Police Academy, had attended the FBI criminal-investigation school. Before he retired, he was a supervisor at York County Prison.

He testified in a low, mildly surly voice, a whine of self-pity always present underneath. He was unashamedly ignorant and utterly devoid of curiosity. He believed, he stated, in a literal reading of the Book of Genesis. He knew almost nothing about evolution except that "it's happenstance, it just happened," and soon revealed an equal ignorance of intelligent design. "I just know that it's another scientific theory that we thought would be good to have presented to the students."

Worse even than his ignorance were his lies. The most important part of his testimony, and the source of one of the most dramatic moments in court, was his contention that neither he nor board president Alan Bonsell had ever used the word "creationism" in the afore-reported school board meetings. They had been fixed on the scientific theory of intelligent design from the start. Their intent had never been religious. The reporters had lied.

"Now," Harvey countered, "it's your testimony that at neither meeting no one on the board ever mentioned creationism, isn't that right?" "That's true." "You're very clear on that point, correct?" "Absolutely, because it's just something we didn't do."

Harvey asked him if he'd mind looking at exhibit P-145. The Wizard of Oz tapped a few buttons and there was Buckingham being interviewed by a local TV news reporter outside a school board meeting at which the current biology book had just been discussed.

"The book that was presented to me," Buckingham said on the video, "was laced with Darwinism from beginning to end. It's okay to teach Darwin, but you have to balance it with something else, such as creationism."

Buckingham looked both irritated and put-upon, and claimed that "when I was walking from my car to the building, here's this lady and here's a cameraman, and I had on my mind all the newspaper articles saying we were talking about creationism, and I had it in my mind to make sure, make double sure, nobody talks about creationism, we're talking intelligent design. I had it on my mind, I was like a deer in the headlights of a car, and I misspoke. Pure and simple, I made a human mistake."*

During this testimony, if you looked to the back of the court you could see Bonsell, president of the school board, grinning as Buckingham screwed things up. It hardly seemed to matter to him. Their case could not be damaged. God was on their side.

The two local reporters, Heidi Bernhard-Bubb and Joe Maldonado, were called to testify to the truthfulness of their articles. A new lawyer for the defense, Edward White III, came forward to cross-examine them.

White is famous for defending anti-abortion activists who listed doctors' personal information, in the form of "wanted" posters, on an Internet site called the "Nuremberg Files." Three doctors listed on the site were killed in the Nineties, and at one time, I am told, there were "X"s over their faces. The site is now shut down, but if you search the web for Christians of this persuasion you can still find sites listing the names of the three murdered doctors.

White's face was not one within which I could find anything to like. In repose, his head was tilted back in petulant defiance. A superior sneer worked his mouth, and his eyes were arrogant and cold. But he was rarely in repose. Every few minutes, his hand would

---

*Neither Buckingham nor his lawyers could be reached for comment. Later in the trial, the plaintiff's attorneys were able to shed light on what he was trying to hide: namely, that a conscious decision had been made to replace any mention of "creationism" with the phrase "intelligent design." Whether this was Buckingham's idea or Bonsell's—or in fact was suggested by, say, the Thomas More Law Center or The Discovery Institute, a creation-science "think tank"—is anyone's guess. Buckingham did admit on the stand, though, that he had received legal advice from both organizations at or around the time of the board meetings.

reach up to scratch his nose, then readjust his watch, his glasses, the knot of his tie; now a jacket-shrug, a chin-scratch, a neck-scratch, then back to the glasses, the tie, and this cycle would repeat two or three times before he settled. This was not a man at ease in his own skin.

When he cross-examined Bernhard-Bubb, White questioned the accuracy of her note-taking and suggested that since meetings sometimes lasted three hours, she might have missed things while going to the bathroom. He suggested as well that she had reason to distort her articles in order to please her editors.

Maldonado received even harsher treatment. A handsome man in his thirties, half Hispanic, tough looking, hair shaved close to his head, a fashionable goatee on his chin, Maldonado was polite in an almost military fashion—"Yes, sir," and "That is correct"—and indeed it was soon revealed that he had served in the Air Force for almost seven years. Like Bernhard-Bubb, he was only a part-time writer for the York Daily Record. The rest of the time he was the owner-operator of a sandwich shop.

White went through a brief version of his previously described preening ritual, then turned his contemptuous eye on the witness. "Your primary occupation is running the sandwich shop?" Maldonado replied that it was a toss-up between the sandwich shop and his writing. "You don't have any formal training though, correct?" "No, sir." "And freelancing, I know you love to write, but it's also a way to supplement your income, correct?" "That is correct." "And depending on where the article appears in the paper determines the amount of money you're paid per article, right?" "Yes." "So a front-page story gets you about $65?" "$67.50." "And then if it runs on a cover of one of the sections, the local sections, it's about $60?" "$62.50." "And then just your average story is around $50, right?" "Somewhere in that ballpark, yes." "And it is the editors who decide where in the newspaper your stories will run, correct?"

It was apparent where he was going with this line of questioning—

namely, that the York paper was biased against intelligent design, and therefore it was to Maldonado's economic advantage to lie in order to get his stories onto the front page. Objections were raised and sustained, and the line of questioning died.

There was something so moving to me in this exchange—the idea of a man running a sandwich shop and working a double shift as a reporter to "supplement" his income with $50 articles for the local paper—that I decided the very next day to pay him a visit.

PBJJ's, MALDONADO's SANDWICH SHOP, is in the old Central Market in York, one of those cavernous spaces given over to stands selling crafts and bric-a-brac. Joe is rightfully famous for his "Mojo Chicken" sandwich. Hanging above the counter are two American flags. With him that day was the younger of his two sons, fourteen-year-old Jaryid. His older son, Alex, is at Penn State studying meteorology, and there was a jar on the counter for his college fund. Next to this was a book of poems Maldonado had written.

Jaryid had had open-heart surgery when he was seven months old, which caused some developmental delays. A couple of years ago, Maldonado and his wife, Julie, although appreciative of the teacher's efforts, could see he was suffering in regular school. "It was so much for him, it was just overwhelming to go from one subject to the other, and I never got the sense that he was mastering one lesson before he'd move on to the next one." So they took him out. By "supplementing his income" with the sandwich shop in the mornings and reporting in the evenings, Joe is free to devote every afternoon to educating his son.

Not only had Maldonado—the liberal reporter—been in the military ("I'm proud to say I served my country"); he had also spent his first year of higher education at Jerry Falwell's Liberty University.

As a Christian, he had been forced to think a lot about the issues raised by the trial. He told me that his faith was so deeply embedded in him, it was very hard to lose God from the equation. To him, the more significant question was whether intelligent design was "ready-for-prime-time science." He spoke eloquently on the subject and referred to the fact that Darwin had spent over two decades collecting evidence before he presented his theory.

Before I left the shop, I bought a copy of Maldonado's book, which he inscribed, "And the Lord said, 'Let there be . . . ' Where's the science in that? Joe." Later that night, I opened it with some trepidation and discovered that Joe wrote beautiful poems full of yearning and eroticism and a keen sense of sin. It occurred to me that perhaps Ed White had somehow got hold of a copy, and that when he said, "I know you love to write," he was toying with the idea of reading a few poems in court.

I also went to visit Bernhard-Bubb. She lives in the upper apartment of a nice house in York. She has two children, Ulysses and Bronwyn, both below school age. Here the liberal reporter was found to be a practicing Mormon. While studying at Brigham Young, however, she had been in a band, which she described as being a little like Franz Ferdinand. When her son, in order to impress the guest, started to say "Fuckie, fuckie, fuckie, fuckie," she remained unruffled. She was intelligent, funny, likable, and disagreed with her church on such issues as gay marriage.

Things are not what they seem. Or perhaps, more accurately, only on the outer edge do you find the authentic clichés, and when you find them, if you are me, those that you hate often turn out to be more poignant than repellent.

HEATHER GEESEY, a school board member who supported Bonsell and Buckingham, fell squarely into the repellent category, however, without mitigation. I found her the most terrifying of all the witnesses. A woman who seemed to think—against all evidence—that everything she did or said was astonishingly cute and funny, she clearly relished being on the same team as "President Alan," as she referred to Bonsell, and grinned relentlessly throughout.

Cross-examining her was ACLU lawyer Vic Walczak. Vic had the weary but pugnacious demeanor of a man who had devoted his life, for little pay, to defending the Constitution but knew that the only questions he would ever be asked related to the ACLU's defense of NAMBLA (the North American Man Boy Love Association) and the Ku Klux Klan.

He asked Geesey if she supported the teaching of intelligent design. "Yes." "Because it gave a balanced view of evolution?" "Yes." "It presented an alternative theory?" "Yes." "And the policy talks about gaps and problems with evolution?" "Yes." "Yes. You don't know what those gaps and problems refer to, do you?" "No." "But it's good to teach about those gaps and problems?" "That's our mission statement, yes." "But you have no idea what they are?" "It's not my job, no." "Is it fair to say that you didn't know much about intelligent design in October of 2004?" "Yes." "And you didn't know much about the book *Of Pandas and People* either, did you?" "Correct." "So you had never participated in any discussions of the book?" "No." "And you made no effort independently to find out about the book?" " No." . . . "And no one ever explained to you what intelligent design was about." "No." This went on for quite a while, Geesey grinning throughout as if her ignorance was just the cutest thing, until finally, still smiling happily, she stated that she had relied on the curriculum committee—Bill Buckingham and Alan Bonsell—to make the decision. "And do you know whether Mr. Buckingham has a background in science?" "No, I do not." "Do you know that in fact he doesn't have a background in science?" "I don't know. He's law enforcement, so I would assume he had to take something along the way."

So this was the genesis of the whole thing: an auto repairman appointed an OxyContin-addicted biblical literalist without a shred of knowledge to decide which books the kids should learn from, and a woman who had no curiosity about anything, even her own most deeply held beliefs, seconded the whole idea.

And unless one doubted two seemingly decent professional reporters and a host of other witnesses, she would happily lie.

Judge Jones had practiced law for several years before being picked by then Governor Tom Ridge to chair the state liquor-control board. He had thus far been fair and amiable and funny. One day an objection was raised as to the admissibility of a question put to a witness. A long debate followed with lawyers on both sides giving it their all. When Jones finally ruled the question legitimate, and it was asked again, the witness said, "I don't know." "After all that!" said Jones.

On another occasion when a witness was criticizing the press by saying he didn't pay much attention to people who bought "ink by the bucket," Jones caught my eye and raised his eyebrows.

Soon after this, I visited him in chambers, and he proved to be everything he appeared in court—civilized, thoughtful, and funny. He read extensively. He was more than polite, he was courteous, a gentleman, a man who seemed to treat everyone around him with equal respect. When I complimented him on his humor, he smiled briefly and expressed the hope that it helped relax tension, though he tried never to be cruel. As a lawyer he had experienced cruelty from the bench and was determined never to abuse his power in that way.

He never did while I was there, though Geesey seemed to test the limits of his patience. In her deposition, she had said that she could not remember when the words "intelligent design" had first been used at school board meetings. On the stand she was very clear that it was in June. Perhaps sensing trouble, Gillen asked her if there was anything that had come up since her deposition that allowed her to "date with somewhat more precision" when she first heard the term "intelligent design" being used. Geesey explained that what jogged her memory were two letters written to local newspapers, one of which was authored by her.

As she was about to leave the witness stand, Jones stopped her, saying he was confused. "So am I," responded Geesey in typical perky fashion. "Well," said Jones, "it's more important that I'm not confused than you're not confused." He pointed out that neither letter mentioned intelligent design. Eventually, Walczak was able to establish that she'd been shown her letter at her deposition and in fact had been questioned about it rigorously.

When contacted later about whether she had perjured herself on the stand, Geesey insisted that she had told the truth, calling the lawyers' attempts to discredit her "a big old lie."

Alan Bonsell took the stand a short while later. He is a good-looking, gum-chewing man somewhere in his late thirties or early forties. With the relaxed, entitled, slightly contemptuous manner of a politician or an athlete, he had, throughout the trial, which he visited most often in the afternoons, lounged on the uncomfortable pews, arms stretched out behind him, head back, the grin in place, the mouth chewing. He reminded me of George Bush, in that he exuded a confidence unwarranted by the facts. He had a degree in business management from York College, and I often wondered, and never concluded, whether he was a worse ideologue than Buckingham (because smarter) or just a man of similar personal faith trying to reach a managerial compromise between his friend's more extreme views and those of the rest of the board.

He had a habit of repeating the questions asked of him with added emphasis and a slight upward lilt on the last word or two. "Did I ever think about it? I think about a lot of things."

He admitted that his own personal views about the universe were based on the first two chapters of Genesis but said that at no time had he tried to get creationism into the science class. He believed evolution should be taught, but "when they don't include, you know, problems with it or gaps in a theory, I mean, and you teach it, it almost sounds like they're teaching it as fact."

When asked to come up with an example, he said he'd "seen things on different subjects of how bears turn into whales, you know, this was a natural scientific theory, which I just thought was absurd. There's also statistical things that I've read about how the statistical probability of life happening by itself was basically impossible." I couldn't help wondering what the statistical probability was of God's slapping it all together in six days.

One of the mysteries in the case (aside from who created the universe) was who had anonymously donated the sixty copies of *Of Pandas and People* to the school library. At various times, but most importantly in their depositions, both Buckingham and Bonsell claimed they had no idea who this could be. In court, however, Buckingham admitted that he had gone to his church and asked for donations in order to buy them. He had then given the money to Bonsell's father, who had bought the books and given them to the school.

Steve Harvey, who had the plum job of cross-examining Bonsell, now took him back to his deposition. It soon became abundantly clear that Bonsell knew—and had known at his deposition—the exact provenance of the books. He had lied under oath.*

The exact motivation for lying in the first place never became entirely clear to me, but whatever it was, it did not cause the judge to be happy. When Harvey had finished his cross, Judge Jones asked to see Bonsell's deposition, specifically the section about the donation of the books. He then proceeded to grill Bonsell about the inconsistencies: "The specific question was asked to you, sir: 'You have never spoken to anybody else who was involved with the donation?' And your answer was, 'I don't know the other people.' That didn't say, 'who donated.' That said, 'who was involved with the donation' . . . now, you tell me why you didn't say Mr. Buckingham's name."

Bonsell stumbled, and Judge Jones became increasingly irritated. Why, furthermore, had Bonsell's father been used in the transaction at all? No clear answers were forthcoming. Bonsell was obviously rattled. He had come onto the stand for the early part of his testimony chewing gum. That had gone soon after Harvey started in on him. Gone too was the swagger and the backward tilt of the head. He walked rather humbly back to the pews.

Within an hour or so, both the pose and the gum were back.

PERHAPS I'M NAIVE, or perhaps I have forgotten something, but the Christianity I was raised on had a high regard for truth. How then

---

*Bonsell, of course, denied that he had lied about anything. Whether he or Buckingham or Geesey will be charged with perjury remains unknown.

to explain all this lying? Not just the smaller lies—who bought the books? was the word "creationism" used?—but the larger, insistent lies, the distortion of quotes, the denial of evidence.

Might it all indeed come back to Three Mile Island? The fruit of science, after all, is not just knowledge but technology. Is it because our technology has become so dangerous and baffling that knowledge itself must also be feared? Do the ignorant even recognize a distinction between one and the other?

Forsaken in the shadow of those monstrous cooling towers, perhaps Buckingham and Bonsell cannot be blamed for seeking whatever light and dignity is still available to them: belief in a God who loves them individually, God their father. Where we come from is who we are: I will not be mistaken for a Texan; they will not be mistaken for an ape.

One thing I know is that this small crusade in Pennsylvania was not a narrow assault on ninth-grade science education; it was a war on the scientific method and the value of evidence.

What was being said, not just by Buckingham and Bonsell but by the President and countless others, is that when the evidence is overwhelming and you don't like it, ignore it.

What natural selection will ultimately do with all of us remains to be seen, but in the Dover school board election that took place shortly after the trial, it eliminated nearly all those who supported intelligent design. Only Heather Geesey, who was not up for reelection, survived. Bonsell got fewer votes than anyone.

On December 20, 2005, Judge Jones ruled that the defendants' intelligent design policy violated the Establishment Clause of the First Amendment. In a withering 139-page opinion, he found that the goal of the intelligent design movement is religious in nature, that intelligent design is not science and cannot be taught in Dover schools, and that the board's claimed reason for including intelligent design in the curriculum—solely because it was good science—was a "sham." In referring to board members, he used such words as "striking ignorance" and "breathtaking inanity." Additionally, he wrote that Buckingham and Bonsell "had either testified inconsistently, or lied outright under oath on several occasions," and that "It is ironic that several of these individuals who so staunchly and proudly touted their religious convictions in public, would time and again lie to cover their tracks and disguise the real purpose behind the ID policy." Amen.

WHEN I RETURNED to the Comfort Inn on the last day of the trial, I did not know that I had one more treat in store. Sitting outside the hotel was a man named Scott Mehring.

While covering this story, I was in the habit of asking anyone who looked interesting what they thought of the issues being discussed. Generally speaking, the answers were as limited and predictable as the temperature settings in my hotel room, with by far the largest group opting for the "Off" setting: "Don't know, don't care." But I thought I should ask one last person. Perhaps, finally, I'd find someone who had something new to say on the subject.

Mehring, of Mechanicsburg, Pennsylvania, is forty-eight years old, the onetime owner of a business that had something to do with performance cars. He wore a tight leather motorcycle jacket with no visible shirt underneath and had a Rod Stewart haircut. He liked to party, he told me, and was ready to go out and party hard, but because he'd lost his license for various reasons he had no car and his cab had not yet arrived. So, sure, he'd be happy to share his views with me. I took out my recorder.

"If you go back to the Big Bang," he said, speaking rapidly, "the elements, I'm not sure exactly what they actually were, but whatever the elements were—the atom, the neutron, the proton neutron, whatever it was that created the Big Bang—where did that stuff come from? Spontaneous generation is a dead theory—at one time they thought it was true—left a piece of meat on the ground maggots appeared, they thought the maggots came out of the meat, but actually they just came out to eat the food, so you can't say spontaneous generation created it. . . . Now if you believe in physics, you got the eleventh dimension—it's a new theory, the eleventh dimension—and inside the eleventh dimension they say that there's an infinite number of universes. So my take is that if you die on the earth, we just somehow hop over to the eleventh dimension, and hop from universe to universe to universe forever inside the eleventh dimension. So that means the Bible could be right with everlasting life after we die. But, okay, the elements that started the Big Bang, if that was an intelligent designer? Then you've got another complication. If there was, like, one dude somewhere at the very top that created everything? Well, where did he come from? Who created him? And who created the God who created God? It gives me goose bumps. It's a loop, like in computer programming—it's an endless loop."

He paused and shook his head. His cab had arrived.

"If you think about this too much," he concluded, "you can go insane."

ATUL GAWANDE

The Score

FROM *THE NEW YORKER*

Despite advances in care, the mortality rate for mothers and infants during childbirth was shockingly high in the early part of the twentieth century—until Virginia Apgar came up with her revolutionary test. Since then, the mortality rate has dropped. Atul Gawande looks at these developments in the light of how birthing is practiced today.

AT 5 A.M. on a cool Boston morning not long ago, Elizabeth Rourke—thick black-brown hair, pale Irish skin, and forty-one weeks pregnant—reached over and woke her husband, Chris.

"I'm having contractions," she said.

"Are you sure?" he asked.

"I'm sure."

She was a week past her due date, and the pain was deep and vise-like, nothing like the occasional spasms she'd been feeling. It seemed to come out of her lower back and wrap around and seize her whole abdomen. The first spasm woke her out of a sound sleep. Then came a second. And a third.

She was carrying their first child. So far, the pregnancy had gone well, aside from the exhaustion and nausea of the first trimester, when all she felt like doing was lying on the couch watching "Law & Order" reruns ("I can't look at Sam Waterston anymore without feeling kind of ill," she says). An internist who had just finished her residency, she had landed a job at Massachusetts General Hospital a few months before and had managed to work until this day. Rourke and her husband sat up in bed, timing the contractions by the clock on the bedside table. They were seven minutes apart, and they stayed that way for a while.

Rourke called her obstetrician's office at eight-thirty in the morning, when the phones were turned on, but she knew what the people there were going to say: Don't come to the hospital until the contractions are five minutes apart and last at least a minute. "You take the childbirth class, and they drill it into you a million times," she says. "The whole point of childbirth classes, as far as I could tell, is to make sure you keep your butt out of the hospital until you're really in labor."

The nurse asked if the contractions were five minutes apart and lasted more than a minute. No. Had she broken her water? No. Well, she had a "good start." But she should wait to come in.

During her medical training, Rourke had seen about fifty births and had delivered four babies herself. The last one she had seen was in a hospital parking lot.

"The father had called, saying, 'We're delivering! We're coming to the hospital, and she's delivering!' " Rourke says. "So we were in the E.R. and we went running. It was freezing cold. The car came screeching up to the hospital. The door went flying open. And, sure enough, there the mom was. We could see the baby's head. The resident running next to me got there a second before I did, and he puts his arms down, and the baby went—*phhhoom*—straight into his arms in the middle of the parking lot. It was freezing cold outside, and I'll never forget the steam pouring off the baby. It's blue and crying and the steam was pouring off of it. Then we put this tiny little baby on this enormous stretcher and raced it back into the hospital."

Rourke didn't want to deliver in a parking lot. She wanted a nice, normal vaginal delivery. She didn't even want an epidural. "I didn't want to be confined to bed," she says. "I didn't want to be dead from the waist down. I didn't want a urinary catheter to have to be put in. Everything about the epidural was totally unappealing to me." She was not afraid of the pain. Having seen how too many deliveries had gone, she was mainly afraid of losing her ability to control what was done to her.

She had considered hiring a doula—a birthing coach—to stay with her through delivery. There are studies showing that having a doula can lower the likelihood that a mother will end up with a Cesarean section or an epidural. The more she looked into it, however, the more worried she became about being paired with someone annoying. She thought about delivering with a midwife. But, as a doctor, she felt that she would actually have more control working with another doctor.

By midday, her contractions hadn't really speeded up; they were still coming every seven minutes, maybe every six at most. She was finding it increasingly difficult to get comfortable. "The way it felt best was, strangely enough, to be on all fours," she recalls. So she just hung around the house like that—on all fours during the contractions, her husband close by, both of them nervous and giddy about their baby being on the way.

Finally, at four-thirty in the afternoon, the contractions began coming five minutes apart, and they set off in their Jetta, with the infant car seat installed in the back. When they reached the hospital admissions desk, Rourke was ready. The baby was on the way, and she was eager to bring it into the world as nature had intended.

"I wanted no intervention, no doctors, no drugs," she says. "I didn't want any of that stuff. In a perfect world, I wanted to have my baby in a forest bower attended by fairy sprites."

HUMAN BIRTH IS an astonishing natural phenomenon. Carol Burnett once told Bill Cosby how he could understand what the experience was like. "Take your bottom lip," she said, "pull it as far away from your face as you can, and now pull it over your head." The process is a solution to an evolutionary problem: how a mammal can walk upright, which requires a small, fixed, bony pelvis, and also possess a large brain, which entails a baby whose head is too big to fit through that small pelvis. Part of the solution is that, in a sense, all human mothers give birth prematurely. Other mammals are born mature enough to walk and seek food within hours; our newborns are small and helpless for months. Even so, human birth is a feat involving an intricate sequence of events.

First, a mother's pelvis enlarges. Starting in the first trimester, maternal hormones allow the joints holding the four bones of the pelvis together to stretch and loosen. Almost an inch of space is added. Pregnant women sometimes feel the different parts of their pelvis moving when they walk.

Then, when it's time for delivery, the uterus changes. During gestation, it's a snug, rounded, hermetically sealed pouch; during labor it takes on the shape of a funnel. And each contraction pushes the baby's head down through that funnel, into the pelvis. This happens even in paraplegic women; the mother does not have to do anything.

Meanwhile, the cervix—which is, through pregnancy, a rigid, inch-thick cylinder of muscle and connective tissue capping the end of the funnel—softens and relaxes. Pressure from the baby's head gradually stretches the tissue until it is paper-thin—a process known as "effacement." A small circular opening appears, and each contraction widens it, like a tight shirt being pulled over a child's head. Until the contractions pull the cervix open about four inches, or ten centimetres—the diameter of the child's head—the child cannot get out. So the state of the cervix determines when birth will occur. At two or three centimetres of dilation, a mother is still in "early" labor. Delivery is many hours away. At between four and seven centimetres, the contractions grow stronger, and "active" labor has begun. At some point, the amniotic sac breaks under the pressure, and the clear fluid surrounding the fetus gushes out. Contractile force increases further.

At between seven and ten centimetres of cervical dilation, the "transition phase," contractions reach their greatest intensity. The contractions press the baby's head into the vagina and the narrowest part of the pelvis's bony ring. The pelvis is usually wider from side to side than front to back, so it's best if the baby emerges with the temples lined up side to side with the mother's pelvis. The top of the head comes into view. The mother has a mounting urge to push. The head comes out, then the shoulders, and suddenly a breathing, wailing child is born. The umbilical cord is cut. The placenta separates from the uterine lining, and, with a slight tug on the cord and a push from the mother, it is extruded. The uterus spontaneously contracts into a clenched ball of muscle, closing off its bleeding sinuses. Typically, the mother's breasts immediately let down colostrum, the first milk, and the newborn can latch on to feed.

That's if all goes well. At almost any step, though, the process can go wrong. For thousands of years, childbirth was the most common cause of death for young women and infants. There's the risk of hemorrhage. The placenta can tear, or separate, or a portion may remain stuck in the uterus after delivery and then bleed torrentially. Or the uterus may not contract after delivery, so that the raw surfaces and sinuses keep bleeding until the mother dies of blood loss. Sometimes the uterus ruptures during labor.

Infection can set in. Once the water breaks, the chances that bacteria will get into the uterus rise with each passing hour. During the nineteenth century, people started to realize that doctors often spread bacteria, because they examined more infected patients than midwives did and failed to wash their contaminated hands. Bacteria routinely invaded and killed the fetus and, often, the mother with it. Puerperal fever was the leading cause of maternal death in the era before antibiotics. Even today, if a mother doesn't deliver within twenty-four hours after her water breaks, she has a forty-per-cent chance of becoming infected.

The most basic problem is "obstruction of labor"—not being able to get the baby out. The baby may be too big, especially when pregnancy continues beyond the fortieth week. The mother's pelvis may be too small, as was frequently the case when lack of vitamin D and calcium made rickets common. The baby might arrive at the birth canal sideways, with nothing but an arm sticking out. It could be a breech, coming butt first and getting stuck with its legs up on its chest. It could be a footling breech, coming feet first, but then getting wedged at the chest with the arms above the head. It could come out head first but get stuck because the head is turned the wrong way. Sometimes the head makes it out, but the shoulders get stuck behind the pubic bone of the mother's pelvis.

These situations are dangerous. When a baby is stuck, the umbilical cord, the only source of fetal blood and oxygen, eventually becomes trapped or compressed, causing the baby to asphyxiate. Mothers have sometimes labored for astonishing lengths of time, unable to deliver, and died with their child in the process. In 1817, for example, Princess Charlotte of Wales, King George IV's twenty-one-year-old daughter, spent fifty hours in active labor with a nine-pound boy. His head was in a sideways position, and too large for Charlotte's pelvis. When he finally emerged, he was stillborn. Six hours later, Charlotte herself died, from hemorrhagic shock. She was King George's only legitimate child. The throne passed to his brother, and then to his niece—which is how Victoria became queen.

MIDWIVES AND DOCTORS HAD LONG SOUGHT WAYS out of such disasters, and the history of obstetrics is the history of these efforts. The first reliably life-saving invention for mothers was called a crochet, or, in another variation, a cranioclast: a sharp-pointed instrument, often with clawlike hooks, which birth attendants used in desperate situations to perforate and crush a fetus's skull, extract the fetus, and save the mother's life.

Many obstetricians made their names by devising methods to get both mother and baby through an obstructed delivery. In the Lovset maneuver for a breech baby with its arms trapped above the head, you take the baby by the hips and turn it sideways, then reach in, take an upper arm, and sweep it down over the chest and out. If a breech baby's arms are out but the head is trapped, you have the Mauriceau-Smellie-Veit maneuver: you place your finger in the baby's mouth, which allows you to pull forcefully while still controlling the head.

The child with its head out but a shoulder stuck—a "shoulder dystocia"—will asphyxiate within five to seven minutes unless it is freed and delivered. Sometimes sharp downward pressure with a fist just above the mother's pubic bone can dislodge the shoulder; if not, there is the Woods corkscrew maneuver, in which you reach in, grab the baby's posterior shoulder, and push it backward to free the child. With the Rubin maneuver, you grab the anterior shoulder and push it forward toward the baby's chest; and with the McRoberts maneuver you sharply flex the mother's legs up onto her abdomen and so lift her pubic bone off the baby's shoulder. Finally, there is the maneuver that no one wanted to put his name to but that through history has saved many babies' lives: you fracture the clavicles—the collar bones—and pull the baby out.

There are dozens of these maneuvers, and, though they have saved the lives of countless babies, each has a high failure rate. Surgery has been known since ancient times as a way to save a trapped baby. Roman law in the seventh century B.C. forbade burial of an undelivered woman until the child had been cut out, in the hope that the child would survive. In 1614, Pope Paul V issued a similar edict, and ordered that the child be baptized if it was still alive. But Cesarean section on a living mother was considered criminal for much of history, because it almost always killed the mother—through hemorrhage and infection—and her life took precedence over that of the child. (The name "Cesarean" section may have arisen from the tale that Caesar was born of his mother, Aurelia, by an abdominal delivery, but historians regard the story as a myth, since Aurelia lived long after his birth.) Only after the development of anesthesia and antisepsis, in the nineteenth century, and, in the early twentieth century, of a double-layer suturing technique that could stop an opened uterus from hemorrhaging, did Cesarean section become a tenable option. Even then it was held in low repute. And that was because a better option was around: the obstetrical forceps.

The story of the forceps is both extraordinary and disturbing, because it is the story of a life-saving idea that was kept secret for more than a century. The instrument was developed in the seventeenth century by Peter Chamberlen (1560-1631), the first of a long line of French Huguenots who delivered babies in London. It looked like a pair of big metal salad tongs, with two blades shaped to fit snugly around a baby's head and handles that locked together with a single screw in the middle. It let doctors more or less yank stuck babies out and, carefully applied, was the first technique that could save both the baby and the mother. The Chamberlens knew that they were onto something, and they resolved to keep the device a family secret. Whenever they were called in to help a mother in obstructed labor, they ushered everyone else out of the room and covered the mother's lower half with a sheet or a blanket so that even she couldn't see what was going on. They kept the secret of the forceps for three generations. In 1670, Hugh Chamberlen, in the third generation, tried and failed to sell it to the French government. Late in his life, he divulged it to an Amsterdam-based surgeon, Roger van Roonhuysen, who kept the technique within his own family for sixty more years. The secret did not get out until the mid-eighteenth century. Once it did, it gained wide acceptance. At the time of Princess Charlotte's failed delivery, in 1817, her obstetrician, Sir Richard Croft, was widely reviled for failing to use forceps. He shot himself to death not long afterward.

By the early twentieth century, the problems of human birth seemed to have been largely solved. Doctors could avail themselves of a range of measures to insure a safe delivery: antiseptics, the forceps, blood transfusions, a drug (ergot) that could induce labor and contract the uterus after delivery to stop bleeding, and even, in desperate situations, Cesarean section. By the nineteen-thirties, most urban mothers had switched from midwife deliveries at home to physician deliveries in the hospital.

But in 1933 the New York Academy of Medicine published a shocking study of 2,041 maternal deaths in childbirth. At least two-thirds, the investigators found, were preventable. There had been no improvement in death rates for mothers in the preceding two decades; newborn deaths from birth injuries had actually increased. Hospital care brought no advantages; mothers were better off delivering at home. The investigators were appalled to find that many physicians simply didn't know what they were doing: they missed clear signs of hemorrhagic shock and other treatable conditions, violated basic antiseptic standards, tore and infected women with misapplied forceps. The White House followed with a similar national report. Doctors may have had the right tools, but midwives without them did better.

The two reports brought modern obstetrics to a turning point. Specialists in the field had shown extraordinary ingenuity. They had developed the knowledge and instrumentation to solve many problems of child delivery. Yet knowledge and instrumentation had proved grossly insufficient. If obstetrics wasn't to go the way of phrenology or trepanning, it had to come up with a different kind of ingenuity. It had to figure out how to standardize childbirth. And it did.

Three-quarters of a century later, the degree to which birth has been transformed by medicine is astounding and, for some, alarming. Today, electronic fetal-heart-rate monitoring is used in more than ninety per cent of deliveries; intravenous fluids in more than eighty per cent; epidural or spinal anesthesia in three-quarters; medicines to speed up labor (the drug of choice is no longer ergot but Pitocin, a synthetic form of the natural hormone that drives contractions) in half. Thirty per cent of American deliveries are now by Cesarean section, and that proportion continues to rise. Something has happened to the field of obstetrics—and, perhaps irreversibly, to childbirth itself.

AN ADMITTING CLERK LED ELIZABETH ROURKE and her husband into a small triage room. A nurse-midwife timed her contractions—they were indeed five minutes apart—and then did a pelvic examination to see how dilated Rourke was. After twelve hours of regular, painful contractions, Rourke figured that she might be at seven or eight centimetres. Instead, she was at two.

It was disheartening news: her labor was only just starting. The nurse-midwife thought about sending her home, but eventually decided to admit her to the labor floor, a horseshoe of twelve patient rooms strung around a nurses' station. For hospitals, deliveries are a good business. If mothers have a positive experience, they stay loyal to the hospital for years. So the rooms are made to seem as warm and inviting as possible. Each had recessed lighting, decorator window curtains, comfortable chairs for the family, individualized climate control. Rourke's even had a Jacuzzi. She spent the next several hours soaking in the tub, sitting on a rubber birthing ball, or walking the halls—stopping to brace herself with each contraction.

By ten-thirty that night, the contractions were coming every two minutes. The doctor on duty for her obstetrician's group performed a pelvic examination. Her cervix was still only two centimetres dilated: the labor had stalled.

The doctor gave her two options. She could have active labor induced with Pitocin. Or she could go home, rest, and wait for true labor to begin. Rourke did not like the idea of using the drug. So at midnight she and her husband went home.

No sooner was she home than she realized that she had made a mistake. The pain was too much. Chris had conked out on the bed, and she couldn't get through this on her own. She held out for two and a half more hours, just to avoid looking foolish, and then got Chris to drive her back. At 2:43 A.M., the nurse scanned her in again— she was still wearing her bar-coded hospital identification bracelet. The obstetrician reëxamined her. Rourke was nearly four centimetres dilated. She had progressed to active labor.

But at this point she had been having regular contractions for twenty-two hours, and was exhausted from sleeplessness and pain. She tried a narcotic called Nubain to dull the pain, and when that didn't work she broke down and asked for an epidural. An anesthesiologist came in and had her sit on the side of the bed with her back to him. She felt a cold wet swipe of antiseptic along her spine, the pressure of a needle, and a twinge that shot down her leg; the epidural catheter was in. The doctor injected a bolus of local anesthetic into the Silastic tubing, and the pain of the contractions melted away into numbness. Then Rourke's blood pressure dropped—a known side effect of epidural injections. The team poured fluids into her intravenously and gave injections of ephedrine to increase her and her baby's blood pressure. It took fifteen minutes to stabilize her blood pressure. But the monitor showed that the baby's heart rate remained normal the whole time, about a hundred and fifty beats a minute. The team dispersed and finally, around 4 A.M., Rourke fell asleep.

At 6 A.M., the obstetrician returned and, to Rourke's dismay, found her still just four centimetres dilated. Her determination to avoid medical interventions ebbed further, and a Pitocin drip was started. The contractions surged. At 7:30 A.M., she was six centimetres dilated. Rourke was elated.

Dr. Alessandra Peccei took over with the new day, and looked at the whiteboard behind the nursing station where the hourly progress of each room is recorded. On a typical morning, a mother in one room might be pushing; in another, a mother might be having her labor induced with medication; in still another, a mother might be just waiting, her cervix only partially dilated and the baby still high.

Rourke was a "G2P0 41.2 wks pit+ 6/100/-2" on the whiteboard—a mother with two gestations, zero born (Rourke had had a miscarriage), forty-one weeks and two days pregnant. She was on Pitocin. Her cervix was six centimetres dilated and a hundred per cent effaced. The baby was at negative-two station, which is about seven centimetres from crowning, that is, from becoming visible.

Peccei went into Rourke's room and introduced herself as the attending obstetrician. Peccei, who was forty-two years old and had delivered more than two thousand babies, projected a comforting combination of competence and friendliness. She had given birth to her own children with a midwife. Rourke felt that they understood one another.

Peccei waited three hours to allow Rourke's labor to progress. At 10:30 A.M., she reëxamined her, and frowned. The cervix was still six centimetres dilated. The baby had not come down any further. Peccei felt along the top of the baby's head for the soft spot in back to get a sense of which way it was facing, and found it facing sideways. The baby was stuck.

Sometimes increasing the strength of the contractions can turn the baby's head in the right direction and push it along. So, using a gloved finger, Peccei punctured the bulging membrane of Rourke's amniotic sac. The waters burst, and immediately the contractions picked up strength and speed, but the baby did not budge. Worse, its heart rate began to drop with each contraction—120, 100, 80 the monitor went, taking almost a minute before returning to normal. It's not always clear what dips like these mean. Malpractice lawyers like to say that they are a baby's "cry for help." In some cases, they are. An abnormal tracing can signal that a baby is getting an inadequate supply of oxygen or blood—the baby's cord may be wrapped around its neck or getting squeezed off altogether. But usually, even when the heart rate takes a prolonged dive, lasting well past the end of a contraction, the baby is fine. A drop in heart rate is often simply what happens when a baby's head is squeezed really hard.

Dr. Peccei couldn't be sure which was the case. She turned off the

Pitocin drip, to reduce the strength of the contractions. She gave Rourke, and therefore the baby, extra oxygen by nasal prong. She scratched at the baby's scalp to irritate it and confirmed that the baby's heart rate responded. The heart rate continued to drop during contractions, but it never failed to recover. After twenty-five minutes, the decelerations disappeared.

Now what? Rourke had been in labor for thirty hours, and her baby didn't seem to be going anywhere.

THERE ARE A HUNDRED AND THIRTY MILLION BIRTHS around the world each year, more than four million of them in the United States. No matter what is done, some percentage will end badly. All the same, physicians have long had an abiding faith that they could step in and at least reduce that percentage. When the national reports of the nineteen-thirties proved that obstetrics had failed to do so, and that incompetence was an important reason, the medical profession turned to a strategy of instituting strict regulations on individual practice. Training requirements were established for physicians delivering babies. Hospitals set firm rules about who could do deliveries, what steps they had to follow, and whether they would be permitted to use forceps and other risky interventions. Hospital and state authorities investigated maternal deaths for aberrations from basic standards.

These standards reduced the number of maternal deaths substantially. In the mid-thirties, delivering a child had been the single most dangerous event in a woman's life: one in a hundred and fifty pregnancies ended in the death of the mother. By the fifties, owing in part to the tighter standards, and in part to the discovery of penicillin and other antibiotics, the risk of death for a mother had fallen more than ninety per cent, to just one in two thousand.

But the situation wasn't so encouraging for newborns: one in thirty still died at birth—odds that were scarcely better than those of the century before—and it wasn't clear how that could be changed.

Then a doctor named Virginia Apgar, who was working in New York, had an idea. It was a ridiculously simple idea, but it transformed obstetrics and the nature of childbirth. Apgar was an unlikely revolutionary for obstetrics. For starters, she had never delivered a baby—not as a doctor and not even as a mother.

Apgar was one of the first women to be admitted to the surgical residency at Columbia University College of Physicians and Surgeons, in 1933. The daughter of a Westfield, New Jersey, insurance executive, she was tall and would have been imposing if not for her horn-rimmed glasses and bobby-pinned hair. She had a combination of fearlessness, warmth, and natural enthusiasm that drew people to her. When anyone was having troubles, she would sit down and say, "Tell Momma all about it." At the same time, she was exacting about everything she did. She wasn't just a talented violinist; she also made her own instruments. She began flying single-engine planes at the age of fifty-nine. When she was a resident, a patient she had operated on died after surgery. "Virginia worried and worried that she might have clamped a small but essential artery," L. Stanley James, a colleague of hers, later recalled. "No autopsy permit could be obtained. So she secretly went to the morgue and opened the operative incision to find the cause. That small artery had been clamped. She immediately told the surgeon. She never tried to cover a mistake. She had to know the truth no matter what the cost."

At the end of her surgical residency, her chairman told her that, however good she was, a female surgeon had little chance of attracting patients. He persuaded her to join Columbia's faculty as an anesthesiologist, then a position of far lesser status. She threw herself into the job, and became the second woman in the country to be board-certified in anesthesiology. She established anesthesia as its own division at Columbia and, eventually, as its own department, on an equal footing with surgery. She administered anesthesia to more than twenty thousand patients during her career. She even carried a scalpel and a length of tubing in her purse, in case a passerby needed an emergency airway—and, apparently, employed them successfully more than a dozen times. "Do what is right and do it now," she used to say.

Throughout her career, the work she loved most was providing anesthesia for child deliveries. But she was appalled by the poor care that many newborns received. Babies who were born malformed or too small or just blue and not breathing well were listed as stillborn, placed out of sight, and left to die. They were believed to be too sick to live. Apgar believed otherwise, but she had no authority to challenge the conventions. She was not an obstetrician, and she was a female in a male world. So she took a less direct, but ultimately more powerful, approach: she devised a score.

The Apgar score, as it became known universally, allowed nurses to rate the condition of babies at birth on a scale from zero to ten. An infant got two points if it was pink all over, two for crying, two for taking good, vigorous breaths, two for moving all four limbs, and two if its heart rate was over a hundred. Ten points meant a child born in perfect condition. Four points or less meant a blue, limp baby.

The score was published in 1953, and it transformed child delivery. It turned an intangible and impressionistic clinical concept—the condition of a newly born baby—into a number that people could collect and compare. Using it required observation and documentation of the true condition of every baby. Moreover, even if only because doctors are competitive, it drove them to want to produce better scores—and therefore better outcomes—for the newborns they delivered.

Around the world, virtually every child born in a hospital had an Apgar score recorded at one minute after birth and at five minutes after birth. It quickly became clear that a baby with a terrible Apgar score at one minute could often be resuscitated—with measures like oxygen and warming—to an excellent score at five minutes. Spinal and then epidural anesthesia were found to produce babies with better scores than general anesthesia. Neonatal intensive-care units sprang into existence. Prenatal ultrasound came into use to detect problems for deliveries in advance. Fetal heart monitors became standard. Over the years, hundreds of adjustments in care were made, resulting in what's sometimes called "the obstetrics package." And that package has produced dramatic results. In the United States today, a full-term baby dies in just one out of five hundred child-births, and a mother dies in one in ten thousand. If the statistics of 1940 had persisted, fifteen thousand mothers would have died last year (instead of fewer than five hundred)—and a hundred and twenty thousand newborns (instead of one-sixth that number).

THERE'S A PARADOX HERE. Ask most research physicians how a profession can advance, and they will talk about the model of "evidence-based medicine"—the idea that nothing ought to be intro-duced into practice unless it has been properly tested and proved effective by research centers, preferably through a double-blind, ran-domized controlled trial. But, in a 1978 ranking of medical specialties according to their use of hard evidence from randomized clinical trials, obstetrics came in last. Obstetricians did few randomized trials, and when they did they ignored the results. Careful studies have found that fetal heart monitors provide no added benefit over having nurses simply listen to the baby's heart rate hourly. In fact, their use seems to increase unnecessary Cesarean sections, because slight abnormalities in the tracings make everyone nervous about waiting for vaginal delivery. Nonetheless, they are used in nearly all hospital deliveries. Forceps have virtually disappeared from the delivery wards, even though several studies have compared forceps delivery to Cesarean section and found no advantage for Cesarean section. (A few found that mothers actually did better with forceps.)

Doctors in other fields have always looked down their masked noses on their obstetrical colleagues. Obstetricians used to have trou-ble attracting the top medical students to their specialty, and there seemed little science or sophistication to what they did. Yet almost nothing else in medicine has saved lives on the scale that obstetrics has. Yes, there have been dazzling changes in what we can do to treat disease and improve people's lives. We now have drugs to stop strokes and to treat cancers; we have coronary-artery stents, artificial joints, and mechanical respirators. But those of us in other fields of medicine don't use these measures anywhere near as reliably and as safely as obstetricians use theirs.

Ordinary pneumonia, for instance, remains the fourth most common cause of death in affluent countries, and the death rate has actually worsened in the past quarter century. That's in part because pneumonias have become more severe, but it's also because we doctors haven't performed all that well. Research trials have shown that patients who are hospitalized with pneumonia are less likely to die if the right antibiotics are started within four hours of their arrival. But we pay little attention to what happens in practice. A recent study has shown that forty per cent of pneumonia patients do not get the antibiotics on time. When we do give the antibiotics, twenty per cent of patients get the wrong kind.

In obstetrics, meanwhile, if a strategy seemed worth trying doctors did not wait for research trials to tell them if it was all right. They just went ahead and tried it, then looked to see if results improved. Obstetrics went about improving the same way Toyota and General Electric did: on the fly, but always paying attention to the results and trying to better them. And it worked. Whether all the adjustments and innovations of the obstetrics package are necessary and beneficial may remain unclear—routine fetal heart monitoring is still controversial, for example. But the package as a whole has made child delivery demonstrably safer and safer, and it has done so despite the increasing age, obesity, and consequent health problems of pregnant mothers.

The Apgar score changed everything. It was practical and easy to calculate, and it gave clinicians at the bedside immediate information on how they were doing. In the rest of medicine, we measure dozens of specific things: blood counts, electrolyte levels, heart rates, viral titers. But we have no measure that puts them together to grade how the patient as a whole is faring. It's like knowing, during a basketball game, how many blocked shots and assists and free throws you have had, but not whether you are actually winning. We have only an impression of how we're performing—and sometimes not even that. At the end of an operation, have I given my patient a one-in-fifty chance of death, or a one-in-five-hundred chance? I don't know. I have no feel for the difference along the way. "How did the surgery go?" the patient's family will ask me. "Fine," I can only say.

The Apgar effect wasn't just a matter of giving clinicians a quick objective read of how they had done. The score also changed the choices they made about how to do better. When chiefs of obstetrics services began poring over the Apgar results of their doctors and midwives, they started to think like a bread-factory manager taking stock of how many loaves the bakers burned. They both want solutions that will lift the results of every employee, from the novice to the most experienced. That means sometimes choosing reliability over the possibility of occasional perfection.

The fate of the forceps is a revealing example. I spoke to Dr. Watson Bowes, Jr., an emeritus professor of obstetrics at the University of North Carolina and the author of a widely read textbook chapter on forceps technique. He started practicing in the nineteen-sixties, when fewer than five per cent of deliveries were by C-section and more than forty per cent were with forceps. Yes, he said, many studies did show fabulous results for forceps. But they only showed how well forceps deliveries could go in the hands of highly experienced obstetricians at large hospitals. Meanwhile, the profession was being held responsible for improving Apgar scores and mortality rates for newborns everywhere—at hospitals small and large, with doctors of all levels of experience.

"Forceps deliveries are very difficult to teach—much more difficult than a C-section," Bowes said. "With a C-section, you stand across from the learner. You can see exactly what the person is doing. You can say, 'Not there. *There.*' With the forceps, though, there is a feel that is very hard to teach." Just putting the forceps on a baby's head is tricky. You have to choose the right one for the shape of the mother's pelvis and the size of the child's head—and there are at least half a dozen types of forceps. You have to slide the blades symmetrically along the sides, traveling exactly in the space between the ears and the eyes and over the cheekbones. "For most residents, it took two or three years of training to get this consistently right," he said. Then a doctor must apply forces of both traction and compression—pulling, his chapter explained, with an average of forty to seventy pounds of axial force and five pounds of fetal skull compression. "When you put tension on the forceps, you should have some sense that there is movement." Too much force, and skin can tear, the skull can fracture, a fatal brain hemorrhage may result. "Some residents had a real feel for it," Bowes said. "Others didn't."

The question facing obstetrics was this: Is medicine a craft or an industry? If medicine is a craft, then you focus on teaching obstetricians to acquire a set of artisanal skills—the Woods corkscrew maneuver for the baby with a shoulder stuck, the Lovset maneuver for the breech baby, the feel of a forceps for a baby whose head is too big. You do research to find new techniques. You accept that things will not always work out in everyone's hands.

But if medicine is an industry, responsible for the safest possible delivery of millions of babies each year, then the focus shifts. You seek reliability. You begin to wonder whether forty-two thousand obstetricians in the U.S. could really master all these techniques. You notice the steady reports of terrible forceps injuries to babies and mothers, despite the training that clinicians have received. After Apgar, obstetricians decided that they needed a simpler, more predictable way to intervene when a laboring mother ran into trouble. They found it in the Cesarean section.

JUST AFTER SEVEN-THIRTY, in the thirty-ninth hour of labor, Elizabeth Rourke had surgery to deliver her baby. Dr. Peccei had offered her the option of a Cesarean eight hours before, but Rourke refused. She hadn't been ready to give up on pushing her baby out into the world, and, though the doctor doubted that Rourke's efforts would succeed, the baby was doing fine on the heart monitor. There was no harm in Rourke's continuing to try. The doctor increased the Pitocin dose gradually, until it was as high as the baby's heart rate allowed. Despite the epidural, the contractions became fiercely painful. And there was progress: by 3 P.M., Rourke's cervix had dilated to eight centimetres. The contractions had pushed the baby forward two centimetres. Even Peccei began to think that Rourke might actually make the delivery happen.

Three hours later, however, the baby's head was no lower and was still sideways; Rourke's cervix hadn't dilated any further. When Dr. Peccei offered her a Cesarean again, she accepted, gratefully.

The Pitocin drip was turned off. The contraction monitor was removed. There was just the swift *tock-tock-tock* of the fetal heart monitor. Peccei introduced a colleague who would do the operation—Rourke had been in labor so long she'd gone through three shifts of obstetricians. She was wheeled to a spacious, white-tiled operating room down the hall. Her husband, Chris, put on green scrubs, a tie-on mask, a bouffant surgical cap, and blue booties over his shoes. He took a chair next to her at the head of the operating table and placed his hand on her shoulder. The anesthesiologist put extra medication in her epidural and pricked at the skin of her belly to make sure that the band of numbness was wide enough. The nurse painted her skin with a yellow-brown antiseptic. Then the cutting began.

The Cesarean section is among the strangest operations I have seen. It is also one of the most straightforward. You press a No. 10 blade down through the flesh, along a side-to-side line low on the bulging abdomen. You divide the skin and golden fat with clean, broad strokes. Using a white gauze pad, you stanch the bleeding points, which appear like red blossoms. You slice through the fascia covering the abdominal muscle, a husk-like fibrous sheath, and lift it to reveal the beefy red muscle underneath. The rectus abdominis muscle lies in two vertical belts that you part in the middle like a curtain, metal retractors pulling left and right. You cut through the peritoneum, a thin, almost translucent membrane. Now the uterus—plum-colored, thick, and muscular—gapes into view. You make a small initial opening in the uterus with the scalpel, and then you switch to bandage scissors to open it more swiftly and easily. It's as if you were cutting open a tough, leathery fruit.

Then comes what still seems surreal to me. You reach in, and, instead of finding a tumor or some other abnormality, as surgeons usually do when we go into someone's belly, you find five tiny wiggling toes, a knee, a whole leg. And suddenly you realize that you have a new human being struggling in your hands. You almost forget the mother on the table. The infant can sometimes be hard to get out. If the head is deep in the birth canal, you have to grasp the baby's waist, stand up straight, and *pull*. Sometimes you have to have someone push on the baby's head from below. Then the umbilical cord is cut. The baby is swaddled. The nurse records the Apgar score.

After the next uterine contraction, you deliver the placenta through the wound. With a fresh gauze pad, you wipe the inside of the mother's uterus clean of clots and debris. You sew it closed with two baseball-stitched layers of stout absorbable suture. You sew the muscle fascia back together with another suture, then sew the skin. And you are done.

This procedure, once a rarity, is now commonplace. Whereas before obstetricians learned one technique for a foot dangling out, another for a breech with its arms above its head, yet another for a baby with its head jammed inside the pelvis, all tricky in their own individual ways, now the solution is the same almost regardless of the problem: the C-section. Every obstetrician today is comfortable doing a C-section. The procedure is performed with impressive consistency.

Straightforward as these operations are, they can go wrong. The child can be lacerated. If the placenta separates and the head doesn't come free quickly, the baby can asphyxiate. The mother faces significant risks, too. As a surgeon, I have been called in to help repair bowels that were torn and wounds that split open. Bleeding can be severe.

Wound infections are common. There are increased risks of blood clots and pneumonia. Even without any complication, the recovery is weeks longer and more painful than with vaginal delivery. And, in future pregnancies, mothers can face serious difficulties. The uterine scar has a one-in-two-hundred chance of rupturing in an attempted vaginal delivery. There's a similar risk that a new baby's placenta could attach itself to the scar and cause serious bleeding problems. C-sections are surgery. There is no getting around it.

Yet there's also no getting around C-sections. We have reached the point that, when there's any question of delivery risk, the Cesarean is what clinicians turn to—it's simply the most reliable option. If a mother is carrying a baby more than ten pounds in size, if she's had a C-section before, if the baby is lying sideways or in a breech position, if she has twins, if any number of potentially difficult situations for delivery arise, the standard of care requires that a midwife or an obstetrician at least offer a Cesarean section. Clinicians are increasingly reluctant to take a risk, however small, with natural childbirth.

I asked Dr. Bowes how he would have handled obstructed deliveries like Rourke's back in the sixties. His first recourse, as you'd expect, would have included forceps. He had delivered more than a thousand babies with forceps, he said, with a rate of neonatal injury as good as or better than with Cesarean sections, and a far faster recovery for the mothers. Had Rourke been under his care, the odds are excellent that she could have delivered safely without surgery. But Bowes is a virtuoso of a difficult instrument. When the protocols of his profession changed, so did he. "As a professor, you have to be a role model. You don't want to be the cowboy who goes in to do something that your residents are not going to be able to do," he told me. "And there was always uncertainty." Even he had to worry that, someday, his judgment and skill would fail him.

These were the rules of the factory floor. To discourage the inexpert from using forceps—along with all those eponymous maneuvers— obstetrics had to discourage everyone from using them. When Bowes finished his career, in 1999, he had a twenty-four-per-cent Cesarean rate, just like the rest of his colleagues. He has little doubt that he'd be approaching thirty per cent, like his colleagues today, if he were still practicing.

A measure of how safe Cesareans have become is that there is ferocious but genuine debate about whether a mother in the thirty-ninth week of pregnancy with no special risks should be offered a Cesarean delivery as an alternative to waiting for labor. The idea seems the worst kind of hubris. How could a Cesarean delivery be considered without even trying a natural one? Surgeons don't suggest that healthy people should get their appendixes taken out or that artificial hips might be stronger than the standard-issue ones. Our complication rates for even simple procedures remain distressingly high. Yet in the next decade or so the industrial revolution in obstetrics could make Cesarean delivery consistently safer than the birth process that evolution gave us.

Currently, one out of five hundred babies who are healthy and kicking at thirty-nine weeks dies before or during childbirth—a historically low rate, but obstetricians have reason to believe that scheduled C-sections could avert at least some of these deaths. Many argue that the results for mothers are safe, too. Scheduled C-sections are certainly far less risky than emergency C-sections—procedures done quickly, in dire circumstances, for mothers and babies already in distress. One recent American study has raised concerns about the safety of scheduled C-sections, but two studies, one in Britain and one in Israel, actually found scheduled C-sections to have *lower* maternal mortality than vaginal delivery. Mothers who undergo planned C-sections may also (though this remains largely speculation) have fewer problems later in life with incontinence and uterine prolapse.

And yet there's something disquieting about the fact that childbirth is becoming so readily surgical. Some hospitals are already doing Cesarean sections in more than half of child deliveries. It is not mere nostalgia to find this disturbing. We are losing our connection to yet another natural process of life. And we are seeing the waning of the art of childbirth. The skill required to bring a child in trouble safely through a vaginal delivery, however unevenly distributed, has been nurtured over centuries. In the medical mainstream, it will soon be lost.

Skeptics have noted that Cesarean delivery is suspiciously convenient for obstetricians' schedules and, hour for hour, is paid more handsomely than vaginal birth. Obstetricians say that fear of malpractice suits pushes them to do C-sections more frequently than even they consider necessary. Putting so many mothers through surgery is hardly cause for celebration. But our deep-seated desire to limit risk to babies is the biggest force behind its prevalence; it is the price exacted by the reliability we aspire to.

In a sense, there is a tyranny to the score. Against the score for a newborn child, the mother's pain and blood loss and length of recovery seem to count for little. We have no score for how the mother does, beyond asking whether she lived or not—no measure to prod us to improve results for her, too. Yet this imbalance, at least, can surely be righted. If the child's well-being can be measured, why not the mother's, too? Indeed, we need an Apgar score for everyone who encounters medicine: the psychiatry patient, the patient on the hospital ward, the person going through an operation, and the mother in childbirth. My research group recently came up with a surgical Apgar score—a ten-point surgical rating based on the amount of blood loss, the lowest heart rate, and the lowest blood pressure that a patient experiences during an operation. We still don't know if it's perfect. But all patients deserve a simple measure that indicates how well or badly they have come through—and that pushes the rest of us to innovate.

"I watched, you know," Rourke says. "I could see the whole thing in the surgical lights. I saw her head come out!" Katherine Anne was seven pounds, fifteen ounces at birth, with brown hair, blue-gray eyes, and soft purple welts where her head had been wedged sideways deep inside her mother's pelvis. Her Apgar score was eight at one minute and nine at five minutes—nearly perfect.

Her mother had a harder time. "I was a wreck," Rourke says. "I was so exhausted I was basically stuporous. And I had unbearable pain." She'd gone through almost forty hours of labor and a Cesarean section. Dr. Peccei told her the next morning, "You got whipped two ways, and you are going to be a mess." She was so debilitated that her milk did not come in.

"I felt like a complete failure, like everything I had set out to do I failed to do," Rourke says. "I didn't want the epidural and then I begged for the epidural. I didn't want a C-section, and I consented to a C-section. I wanted to breast-feed the baby, and I utterly failed to breast-feed." She was miserable for a week. "Then one day I realized, 'You know what? This is a stupid thing to think. You have a totally gorgeous little child and it's time to pay more attention to your totally gorgeous little child.' Somehow she let me put all my regrets behind me."

Jennifer Couzin

Truth and Consequences

FROM *SCIENCE*

When a scientist is accused of falsifying data, what happens to that scientist's lab? As Jennifer Couzin reports, the lab's post-docs and young researchers can pay as much of a price as the wrongdoer.

I N THOSE FIRST DISORIENTING MONTHS, as fall last year turned to winter and the sailboats were hauled out of nearby lakes, the graduate students sometimes gathered at the Union Terrace, a popular student hangout. There, they clumped together at one of the brightly colored tables that look north over Lake Mendota, drinking beer and circling endlessly around one agonizing question: What do you do when your professor apparently fakes data, and you are the only ones who know?

Chantal Ly, 32, had already waded through 7 years of a Ph.D. program at the University of Wisconsin (UW), Madison. Turning in her mentor, Ly was certain, meant that "something bad was going to happen to the lab." Another of the six students felt that their adviser, geneticist Elizabeth Goodwin, deserved a second chance and wasn't certain the university would provide it. A third was unable for weeks to believe Goodwin had done anything wrong and was so distressed by the possibility that she refused to examine available evidence.

Two days before winter break, as the moral compass of all six swung in the same direction, they shared their concerns with a university administrator. In late May, a UW investigation reported data falsification in Goodwin's past grant applications and raised questions about some of her papers. The case has since been referred to the federal Office of Research Integrity (ORI) in Washington, D.C. Goodwin, maintaining her innocence, resigned from the university at the end of February. (Through her attorney, Goodwin declined to comment for this story.)

Although the university handled the case by the book, the graduate students caught in the middle have found that for all the talk about honesty's place in science, little good has come to them. Three of the students, who had invested a combined 16 years in obtaining their Ph.D.s, have quit school. Two others are starting over, one moving to a lab at the University of Colorado, extending the amount of time it will take them to get their doctorates by years. The five graduate students who spoke with *Science* also described discouraging encounters with other faculty members, whom they say sided with Goodwin before all the facts became available.

Fraud investigators acknowledge that outcomes like these are typical. "My feeling is it's never a good career move to become a whistle-blower," says Kay Fields, a scientific investigator for ORI, who depends on precisely this occurrence for misconduct cases to come to light. ORI officials estimate that between a third and half of nonclinical misconduct cases—those involving basic scientific research—are brought by postdoctoral fellows or graduate students like those in Goodwin's lab. And the ones who come forward, admits ORI's John

Dahlberg, often suffer a "loss of time, loss of prestige, [and a] loss of credibility of your publications."

Indeed, Goodwin's graduate students spent long hours debating how a decision to alert administrators might unravel. Sarah LaMartina, 29, who gravitated to biology after its appeal outshone her childhood plan to become a veterinarian, had already spent 6 years in graduate school and worried whether all that time and effort would go to waste. "We kept thinking, 'Are we just stupid [to turn Goodwin in]?'" says LaMartina, whose midwestern accent reflects her Wisconsin roots. "Sure, it's the right thing to do, but right for who? . . . Who is going to benefit from this? Nobody."

GOODWIN, IN HER LATE 40S, had come to the University of Wisconsin in 2000 from Northwestern University in Chicago, Illinois, and was awarded tenure by UW soon after. Landing in Wisconsin was something of a homecoming for her; she had done a postdoc under Judith Kimble, a prominent developmental geneticist in the same department. Goodwin studied sex determination in worms during their early development and has published more than 20 papers on that and other subjects in various prominent journals (including, in 2003, *Science*). Goodwin was also the oldest of a crop of female faculty members hired in recent years by genetics department chair Michael Culbertson. "She was the role model," he says.

In the beginning, the Goodwin lab had a spark. Students recall being swept up in its leader's enthusiasm when, seeking a lab in which to settle, they rotated through for a month during their first year of graduate school. Goodwin pushed her students to believe that compelling scientific results were always possible, boosting their spirits during the low points that invariably strike Ph.D. hopefuls. She held annual Christmas parties at her home west of Madison. Once, she took the entire lab on a horseback-riding trip.

Then, last October, everything changed. One afternoon, in the conference room down the hall from the lab, Ly told Goodwin she was concerned about her progress: The project she'd been working on, Ly felt, wasn't yielding usable results. Despite months of effort, Ly was unable to replicate earlier observations from the lab.

"At that time, she gave me three pages of a grant [application]," Ly recalled recently. The proposal, which was under review at the National Institutes of Health (NIH), sought to broaden a worm genetics project that another student, third-year Garett Padilla, had begun. Goodwin, Ly says, told her that the project, on a new, developmentally important worm gene, was "really promising, but there's so many aspects of it there's no way he can work on everything." Goodwin urged Ly to peruse the pages and see whether the gene might interest her as a new project.

Reading the grant application set off alarm bells for Ly. One figure, she quickly noticed, was represented as unpublished data even though it had appeared in a 2004 paper published by Goodwin's lab.

Ly and Padilla sat back to back at desks in the corridor outside the lab. When she showed him the pages from the grant application, he too was shaken. "There was one experiment that I had just not done," as well as several published and unpublished figures that seemed to have been manipulated, he says. Two images apparently identical to those already published were presented as unpublished and as representing proteins different from the published versions. "I remember being overwhelmed and not being able to deal with it at that moment," says Padilla.

A bearish 25-year-old with a closely cropped beard and wire-rimmed glasses, Padilla speaks softly, with deliberation. Bored by bench work, he was considering leaving biology research for law school and had discussed the possibility with Goodwin. She had urged him to "stick it out," he says. "Everybody goes through a phase where they don't want to be here," he recalls Goodwin telling him.

At a loss after seeing the grant application, Padilla consulted two scientists for advice: his fiancée's adviser, a physiology professor at the university, and Scott Kuersten, a former postdoc in Goodwin's lab who had been dating LaMartina for several years and who happened to be in town. Kuersten and Padilla talked for about an hour and together examined the pages of the proposal. Kuersten, now at Ambion, a biotechnology company in Austin, Texas, advised Padilla to ask Goodwin for an explanation, as did the physiologist.

Padilla steeled himself for a confrontation. On Halloween day, he paced nervously outside Goodwin's office, summoning the courage to knock. The conversation did not go well, says Padilla.

In a computer log of events he had begun to keep at Kuersten's urging, which he shared with *Science*, Padilla wrote that Goodwin denied lifting a Western blot image from a published paper and presenting it as unpublished work, although, he added in the log, "She became extremely nervous and repeatedly said, 'I fucked up.'" Padilla also noted: "I left feeling that no issues were resolved." His confusion deepened when Goodwin later that day blamed the problem on a computer file mix-up.

Meanwhile, word was leaking out to others in the lab that something was terribly wrong. Two days later, Padilla called a meeting of all current lab members: six graduate students and the lab technician. To ensure privacy, the group, minus Ly, who had recently had a baby girl, convened in the nearby engineering library. Padilla laid out the grant papers for all to see.

In that meeting, ensconced in the library, the grad students hesitated at the thought of speaking with the administration. "We had no idea what would happen to us, we had no idea what would happen to Betsy, we had no idea how the university would react," says LaMartina, who admits to some distrust of authority and also a belief that people who err deserve a second chance.

Ly felt less charitable toward Goodwin but confesses that at first she considered only her own predicament. In many ways, just reaching graduate school was a triumph for Ly, and she badly wanted that doctorate. In 1981, when Ly was 8 years old, her family fled Cambodia for the Chicago suburbs. Around Ly's neck hangs a goldplated French coin, a 20-franc piece her curator father had collected before he was killed in his country's civil war.

In Chicago, Ly's mother worked long hours and put her daughter through Wellesley College in Massachusetts. When Ly moved to Madison, so did her husband, now an anesthesia resident, and her mother, who speaks little English and cannot drive. "Here I am, I've invested so much time in grad school, and this happens. If we let someone know . . . " she says, her voice trailing off.

The students decided that Padilla needed to speak with Goodwin a second time, in hope of extracting a clear account of what went wrong or even a retraction of the grant application. Four days after his first nerve-wracking encounter, Padilla was in Goodwin's office again. This time, the conversation put him at ease. Padilla says Goodwin asked for forgiveness and praised him for, as he wrote in the log, "pushing this issue." She told him that the grant application was unlikely to be funded—an assertion that turned out to be untrue given that NIH approved it—but offered to e-mail her NIH contact citing some of the problems in the application. Goodwin subsequently sent that e-mail, on which Padilla was copied. He left the encounter relieved.

"At that point, I was pretty content to leave it alone," he says. "I felt like we had compromised on a resolution."

ANOTHER STUDENT, HOWEVER, was finding little peace. Mary Allen, 25 and in her fourth year of graduate school, couldn't shake a sense of torment about what her mentor might have done. A bookworm who squeezed 3 years of high school into one and entered college at age 15, Allen is guided by unambiguous morals and deep religious convictions, attending a local church regularly and leading a youth group there. She could not fathom that Goodwin had falsified data; at one point, Allen refused even to examine another suspect grant application. But, concerned because Goodwin seemed to have admitted to some wrongdoing, Allen felt she needed to switch labs.

Allen alerted Goodwin that she would likely be moving on. Their mentor then began offering additional explanations for the grant application, say Allen and the others. Goodwin told them that she had mixed up some files and asserted that the files had come to her unlabeled. In a private conversation with Allen, she adamantly denied faking data.

As November wore on, the lab's atmosphere grew ever more stressful and surreal. When Goodwin was present, she chatted with the students about their worm experiments and their families—the same conversations they'd always had.

Yet the strain was taking its toll. LaMartina's appetite declined, and she began losing weight, shedding 15 pounds before the ordeal was over. Padilla called former postdoc Kuersten nearly weekly for advice, and the students talked obsessively with one another. Careful to maintain confidentiality, "the only people we could bounce ideas and solutions off of were each other," says Padilla. The tension even penetrated Goodwin's annual Christmas party. For the first time, several lab members didn't show up.

Deeply worried about how speaking with administrators might impact the more senior students, lab members chose not to alert the university unless the desire to do so was unanimous. Gradually all, including Ly and LaMartina, the most senior among them, agreed that their mentor's denials left them uncomfortable and concerned that she might falsify data in the future. "My biggest worry was what if we didn't turn her in . . . and different grad students got stuck in our position," says Allen.

Two days before exams ended, on 21 December, Ly and Padilla met together with Culbertson and showed him the suspect grant pages. Culbertson didn't know what to think at first, he says, but "when somebody comes to me with something like that, I have to investigate."

CULBERTSON QUICKLY REFERRED THE MATTER to two university deans, who launched an informal inquiry to determine whether a more formal investigation was warranted. As is customary,

Goodwin remained on staff at the university during this time. She vigorously denied the charges against her, telling Culbertson and the students in a joint meeting that the figures in question were place-holders she had forgotten to swap out. According to Padilla's log of that meeting, Goodwin explained that she "was juggling too many commitments at once" when the proposal was submitted.

Two biology professors ran the informal inquiry, conducting inter-views with Goodwin and her students. One of the two, Irwin Gold-man, was also a dean, and he became the students' unofficial therapist and news source. At their first meeting in January, Goldman reas-sured the six that their salaries would continue uninterrupted.

The informal inquiry wrapped up a few weeks later, endorsing a more formal investigation. Three university deans, including Gold-man, appointed three faculty scientists to the task.

At about this time, says Goldman, the university grew uneasy about possible fraud not only in the first grant application that the students had seen but also in two others that had garnered funding, from NIH and the U.S. Department of Agriculture. The school can-celed all three grants. After a panicky 2 weeks during which the lab went unfunded, Goldman drew on money from both the college of agricultural and life sciences and the medical school. (Goodwin had a joint appointment at the two.) The students peppered Goldman regularly with questions, seeking advice on whether to talk to a local reporter or how their funding might shake out.

Still, because privacy rules prevented sharing the details, "we felt isolated up on our floor," says Padilla. "There were faculty nearby, but they didn't really know what was going on." Goodwin, meanwhile, all but disappeared from the lab, appearing only once or twice after the investigation began. The students tried to keep up with their projects as they'd always done. They held lab meetings alone before being invited to weekly gatherings with geneticist Philip Anderson's lab.

Most faculty members were aware that an investigation had been launched, and some had heard that Goodwin's students were the informers. That led to disheartening exchanges. A faculty member, asked by one of the students whether they'd done the right thing, told her he didn't know. Rumors reached the students that Goodwin had had "to fake something because her students couldn't produce enough data," says Ly.

In late February, Goodwin resigned. The students say they learned of her departure from a biologist who worked in a neighboring lab.

Three months later, the university released its investigation report, which described "evidence of deliberate falsification" in the three applications for the cancelled grants, totaling $1.8 million in federal funds. In the school's report, which university officials shared with *Science*, investigators also raised questions about three published papers, in *Nature Structural and Molecular Biology, Developmental Biology,* and *Molecular Cell.*

None has been retracted or corrected so far. "We are considering the implications" of the university report, said Lynne Herndon, president and CEO of Cell Press, which publishes *Molecular Cell,* in a statement. The editor of *Nature Structural and Molecular Biology* said she was awaiting the results of the ORI investigation, and the other authors of the *Developmental Biology* paper are reviewing the relevant data, says the journal's editor in chief, Robb Krumlauf of the Stowers Institute for Medical Research in Kansas City, Missouri.

The university investigators also noted other problems in the Goodwin lab. "It appears from the testimony of her graduate students that Dr. Goodwin's mentoring of her graduate students included behaviors that could be considered scientific misconduct—namely, pressuring students to conceal research results that disagreed with desired outcomes and urging them to over-interpret data that the students themselves considered to be preliminary and weak," they wrote in their report.

Goodwin's lawyer in Madison, Dean Strang, disputes the reliability of the school's report. The investigation was "designed under the applicable UW rules to be an informal screening proceeding," and, because Goodwin resigned, "there was no adjudicative proceeding at the administrative level or elsewhere," Strang wrote in an e-mail message. He added that "there are no problems with the three published papers cited in the report (or any others)." Strang declined to address whether Goodwin pressed students to overinterpret data. "Dr. Goodwin will not respond at all to assertions of students in this forum," he wrote.

Culbertson distributed the investigating committee's report to all department faculty members; it even appeared on Madison's evening news. Still, the rapprochement some of the students had hoped for never materialized. "No one ever came up and said, 'I'm sorry,' " Padilla says.

As the graduate students contemplated their futures this spring, they did have one point in their favor: Ironically enough, the sluggish pace of their projects meant that almost none had co-authored papers with Goodwin. But when several of them sat down with their thesis committees to assess their futures, the prognosis was grim. Only one student of the six, who did not reply to *Science*'s request for an interview, was permitted to continue with her original project. She has moved to another Wisconsin lab and hopes to complete her Ph.D. within about a year, according to the others.

Thesis committees and faculty members told Ly, LaMartina, and fourth-year Jacque Baca, 27, that much of their work from Goodwin's lab was not usable and recommended that they start over with a new doctoral project. The reason wasn't necessarily data fraud, the students say, but rather Goodwin's relentless optimism that some now believe kept them clinging to questionable results. Allen, for example, says she sometimes argued but gave in to Goodwin's suggestions that she stick with molecular data Allen considered of dubious quality or steer clear of performing studies that might guard against bias. Ly, on her third, floundering project, says, "I thought I was doing something wrong experimentally that I couldn't repeat these things."

Despite her setback, Baca has chosen to stay at Wisconsin. "It's kind of hard to say" how much time she'll lose, says Baca, who notes that her thesis committee was supportive in helping her find a new lab.

The other four—Ly, LaMartina, Padilla, and Allen—have scattered. Only Allen plans on finishing her Ph.D. Determined to leave Wisconsin behind, she relocated in late March to the University of Colorado, Boulder, where she hopes to start fresh. Members of her church, her husband, and her parents persuaded her to stay in science, which she adores, but she still wonders about the future. "We unintentionally suffer the consequences" of turning Goodwin in, Allen says, noting that it will now take her 8 or 9 years in all to finish graduate school. To her husband's disappointment, their plans for having children have been deferred, as Allen always wanted to wait until she had completed her degree.

For Padilla, the experience cemented the pull of the law. In late July, a month after his wedding, he and his wife moved to Minneapolis-St. Paul, Minnesota, not far from where Padilla grew up, because his wife's adviser, the physiologist, had shifted his lab there. Padilla began law school in the city last week.

LaMartina spent 2 months in a different Wisconsin genetics lab, laboring over a new worm project she'd recently started under Goodwin. That project, however, fell apart in June. She then spent 3 weeks in Seattle and Alaska with Kuersten. During the trip, LaMartina abandoned her Ph.D. plans, and in July, she left Wisconsin for Texas, joining Kuersten at Ambion as a lab technician.

When Ly learned from her thesis committee that her years in the Goodwin lab had come to naught, she left the program and, as a stopgap, joined a cancer lab as a technician. "I decided that I had put my life on hold long enough," Ly says. She intends to leave science altogether and is considering business school.

For Goldman, the dean who supported the graduate students, the experience was bittersweet. Impressed by the students' professionalism and grace under trying circumstances, he came to believe strongly that science needs individuals like them. And although he admits that it's "horrible" that so many of the students were told to start over, "I don't see us changing our standards in terms of what a Ph.D. means," he says.

Still, Goldman does plan to craft formal policies for students who might encounter this situation in the future. The policies, he says, would guarantee that the university protects students from retribution and that their funding remains secure. He hopes that codifying such safeguards will offer potential whistleblowers peace of mind.

In a building with a lobby graced by a fountain shaped like DNA, the Goodwin lab now sits deserted on the second floor. Incubators, pipettes, and empty plastic shoeboxes that once held worms litter its counters. Ly's original fear months before, that something bad would happen to the lab, had proved more prescient than she had imagined.

LAWRENCE K. ALTMAN

The Man on the Table Was 97, But He Devised the Surgery

FROM THE *NEW YORK TIMES*

When the legendary heart surgeon Michael DeBakey was admitted to the hospital with a heart ailment, doctors—his colleagues—had to decide whether, at his advanced age, he would survive the operation he himself created. Lawrence Altman reports on this great medical pioneer's most unexpected milestone.

I N LATE AFTERNOON last Dec. 31, Dr. Michael E. DeBakey, then 97, was alone at home in Houston in his study preparing a lecture when a sharp pain ripped through his upper chest and between his shoulder blades, then moved into his neck.

Dr. DeBakey, one of the most influential heart surgeons in history, assumed his heart would stop in a few seconds.

"It never occurred to me to call 911 or my physician," Dr. DeBakey said, adding: "As foolish as it may appear, you are, in a sense, a prisoner of the pain, which was intolerable. You're thinking, What could I do to relieve myself of it? If it becomes intense enough, you're perfectly willing to accept cardiac arrest as a possible way of getting rid of the pain."

But when his heart kept beating, Dr. DeBakey suspected that he was not having a heart attack. As he sat alone, he decided that a ballooning had probably weakened the aorta, the main artery leading from the heart, and that the inner lining of the artery had torn, known as a dissecting aortic aneurysm.

No one in the world was more qualified to make that diagnosis than Dr. DeBakey because, as a younger man, he devised the operation to repair such torn aortas, a condition virtually always fatal. The operation has been performed at least 10,000 times around the world and is among the most demanding for surgeons and patients.

Over the past 60 years, Dr. DeBakey has changed the way heart surgery is performed. He was one of the first to perform coronary bypass operations. He trained generations of surgeons at the Baylor College of Medicine; operated on more than 60,000 patients; and in 1996 was summoned to Moscow by Boris Yeltsin, then the president of Russia, to aid in his quintuple heart bypass operation.

Now Dr. DeBakey is making history in a different way—as a patient. He was released from Methodist Hospital in Houston in September and is back at work. At 98, he is the oldest survivor of his own operation, proving that a healthy man of his age could endure it.

"He's probably right out there at the cutting edge of a whole generation of people in their 90s who are going to survive" after such medical ordeals, one of his doctors, Dr. James L. Pool, said.

But beyond the medical advances, Dr. DeBakey's story is emblematic of the difficulties that often accompany care at the end of life. It is a story of debates over how far to go in treating someone so old, late-night disputes among specialists about what the patient would want, and risky decisions that, while still being argued over, clearly saved Dr. DeBakey's life.

It is also a story of Dr. DeBakey himself, a strong-willed pioneer who at one point was willing to die, concedes he was at times in denial about how sick he was and is now plowing into life with as much zest and verve as ever.

But Dr. DeBakey's rescue almost never happened.

He refused to be admitted to a hospital until late January. As his health deteriorated and he became unresponsive in the hospital in early February, his surgical partner of 40 years, Dr. George P. Noon, decided an operation was the only way to save his life. But the hospital's anesthesiologists refused to put Dr. DeBakey to sleep because such an operation had never been performed on someone his age and in his condition. Also, they said Dr. DeBakey had signed a directive that forbade surgery.

As the hospital's ethics committee debated in a late-night emergency meeting on the 12th floor of Methodist Hospital, Dr. DeBakey's wife, Katrin, barged in to demand that the operation begin immediately.

In the end, the ethics committee approved the operation; an anesthesiology colleague of Dr. DeBakey's, who now works at a different hospital, agreed to put him to sleep; and the seven-hour operation began shortly before midnight on Feb. 9. "It is a miracle," Dr. DeBakey said as he sat eating dinner in a Houston restaurant recently. "I really should not be here."

The costs of Dr. DeBakey's care easily exceeded $1 million. Methodist Hospital and his doctors say they have not charged Dr. DeBakey. His hospitalizations were under pseudonyms to help protect his privacy, which could make collecting insurance difficult. Methodist Hospital declined to say what the costs were or discuss the case further. Dr. DeBakey says he thinks the hospital should not have been secretive about his illness.

Dr. DeBakey's doctors acknowledge that he got an unusually high level of care. But they said that they always tried to abide by a family's wishes and that they would perform the procedure on any patient regardless of age, if the patient's overall health was otherwise good.

Dr. DeBakey agreed to talk, and permitted his doctors to talk, because of a professional relationship of decades with this reporter, who is also a physician, and because he wanted to set the record straight for the public about what happened and explain how a man nearly 100 years old could survive.

As Dr. DeBakey lay on the couch alone that night, last New Year's Eve, he reasoned that a heart attack was unlikely because periodic checkups had never indicated he was at risk. An aortic dissection was more likely because of the pain, even though there was no hint of that problem in a routine echocardiogram a few weeks earlier.

Mrs. DeBakey and their daughter, Olga, had left for the beach in Galveston, but turned back because of heavy traffic. They arrived home to find Dr. DeBakey lying on the couch. Not wanting to alarm them, he lied and said he had fallen asleep and awakened with a pulled muscle.

"I did not want Katrin to be aware of my self-diagnosis because, in a sense, I would be telling her that I am going to die soon," he said.

An anxious Mrs. DeBakey called two of her husband's colleagues: Dr. Mohammed Attar, his longtime physician, and Dr. Matthias Loebe, who was covering for Dr. Noon. They came to the house quickly and became concerned because Dr. DeBakey had been in excellent health. After listening to him give a more frank account of his pain, they shared his suspicion of an aortic dissection.

Dr. DeBakey and his doctors agreed that for a firm diagnosis he would need a CT scan and other imaging tests, but he delayed them until Jan. 3.

The tests showed that Dr. DeBakey had a type 2 dissecting aortic aneurysm, according to a standard classification system he himself devised years earlier. Rarely did anyone survive that without surgery.

Still, Dr. DeBakey says that he refused admission to Methodist Hospital, in part because he did not want to be confined and he "was hopeful that this was not as bad as I first thought." He feared the operation that he had developed to treat this condition might, at his age, leave him mentally or physically crippled. "I'd rather die," he said.

Over the years, he had performed anatomically perfect operations on some patients who nevertheless died or survived with major complications. "I was trying to avoid all that," he said.

Instead, he gambled on long odds that his damaged aorta would heal on its own. He chose to receive care at home. For more than three weeks, doctors made frequent house calls to make sure his blood pressure was low enough to prevent the aorta from rupturing. Around the clock, nurses monitored his food and drink. Periodically, he went to Methodist Hospital for imaging tests to measure the aneurysm's size.

On Jan. 6, he insisted on giving the lecture he had been preparing on New Year's Eve to the Academy of Medicine, Engineering and Science of Texas, of which he is a founding member. The audience in Houston included Nobel Prize winners and Senator Kay Bailey Hutchison.

Mrs. DeBakey stationed people around the podium to catch her husband if he slumped. Dr. DeBakey looked gray and spoke softly, but finished without incident. Then he listened to another lecture— which, by coincidence, was about the lethal dangers of dissecting aneurysms.

Dr. DeBakey, a master politician, said he could not pass up a chance to chat with the senator. He attended the academy luncheon and then went home.

In providing the extraordinary home care, the doctors were respecting the wishes of Dr. DeBakey and their actions reflected their awe of his power.

"People are very scared of him around here," said Dr. Loebe, the heart surgeon who came to Dr. DeBakey's home on New Year's Eve. "He is the authority. It is very difficult to stand up and tell him what to do."

But as time went on, the doctors could not adequately control Dr. DeBakey's blood pressure. His nutrition was poor. He became short of breath. His kidneys failed. Fluid collected in the pericardial sac covering his heart, suggesting the aneurysm was leaking.

Dr. DeBakey now says that he was in denial. He did not admit to himself that he was getting worse. But on Jan. 23, he yielded and was admitted to the hospital.

Tests showed that the aneurysm was enlarging dangerously; the diameter increased to 6.6 centimeters on Jan. 28, up from 5.2 centimeters on Jan. 3. Dr. Noon said that when he and other doctors showed Dr. DeBakey the scans and recommended surgery, Dr. DeBakey said he would re-evaluate the situation in a few days.

By Feb. 9, with the aneurysm up to 7.5 centimeters and Dr. DeBakey unresponsive and near death, a decision had to be made.

"If we didn't operate on him that day that was it, he was gone for sure," Dr. Noon said.

At that point, Dr. DeBakey was unable to speak for himself. The surgeons gathered and decided they should proceed, despite the dangers. "We were doing what we thought was right," Dr. Noon said, adding that "nothing made him a hopeless candidate for the operation except for being 97." All family members agreed to the operation.

Dr. Bobby R. Alford, one of Dr. DeBakey's physicians and a successor as chancellor of Baylor College of Medicine, said the doctors had qualms. "We could have walked away," he said.

He and Dr. Noon discussed the decision several times. "We recognized the condemnation that could occur," Dr. Alford said. "The whole surgical world would come down on us for doing something stupid, which it might have seemed to people who were not there."

Surgery would be enormously risky and unlikely to offer clear-cut results—either a full recovery or death, Dr. Noon and his colleagues told Mrs. DeBakey, Olga, sons from a first marriage, and Dr. DeBakey's sisters, Lois and Selma. The doctors said Dr. DeBakey might develop new ailments and need dialysis and a tracheostomy to help his breathing. They said the family's decision could inflict prolonged suffering for all involved.

Olga and she "prayed a lot," said Mrs. DeBakey, who is from Germany. "We had a healer in Europe who advised us that he will come through it. That helped us."

Then things got more complicated.

AT THAT POINT the Methodist Hospital anesthesiologists adamantly refused to accept Dr. DeBakey as a patient. They cited a standard form he had signed directing that he not be resuscitated if his heart stopped and a note in the chart saying he did not want surgery for the aortic dissection and aneurysm. They were concerned about his age and precarious physical condition.

Dr. Alford, the 72-year-old chancellor, said he was stunned by the refusal, an action he had never seen or heard about in his career.

Dr. Noon said none of the anesthesiologists had been involved in Dr. DeBakey's care, yet they made a decision based on grapevine information without reading his medical records. So he insisted that the anesthesiologists state their objections directly to the DeBakey family.

Mrs. DeBakey said the anesthesiologists feared that Dr. DeBakey would die on the operating table and did not want to become known as the doctors who killed him. Dr. Joseph J. Naples, the hospital's chief anesthesiologist, did not return repeated telephone calls to his office for comment.

Around 7 p.m., Mrs. DeBakey called Dr. Salwa A. Shenaq, an anesthesiologist friend who had worked with Dr. DeBakey for 22 years at Methodist Hospital and who now works at the nearby Michael E. DeBakey Veterans Affairs Medical Center.

Dr. Shenaq rushed from home. When she arrived, she said, Dr. Naples told her that he and his staff would not administer anesthesia to Dr. DeBakey. She said that a medical staff officer, whom she declined to name, warned her that she could be charged with assault if she touched Dr. DeBakey. The officer also told Dr. Shenaq that she could not give Dr. DeBakey anesthesia because she did not have Methodist Hospital privileges. She made it clear that she did, she said.

Administrators, lawyers and doctors discussed the situation, in particular the ambiguities of Dr. DeBakey's wishes. Yes, Dr. Pool had written on his chart that Dr. DeBakey said he did not want surgery for a dissection. But Dr. Noon and the family thought the note in the chart no longer applied because Dr. DeBakey's condition had so deteriorated and his only hope was his own procedure.

"They were going back and forth," Dr. Shenaq said. "One time, they told me go ahead. Then, no, we cannot go ahead."

To fulfill its legal responsibilities, Methodist Hospital summoned members of its ethics committee, who arrived in an hour. They met with Dr. DeBakey's doctors in a private dining room a few yards from Dr. DeBakey's room, according to five of his doctors who were present.

Their patient was a man who had always been in command. Now an unresponsive Dr. DeBakey had no control over his own destiny.

The ethics committee representatives wanted to follow Texas law, which, in part, requires assurance that doctors respect patient and family wishes.

Each of Dr. DeBakey's doctors had worked with him for more than 20 years. One, Dr. Pool, said they felt they knew Dr. DeBakey well enough to answer another crucial question from the ethics committee: As his physicians, what did they believe he would choose for himself in such a dire circumstance if he had the ability to make that decision?

Dr. Noon said that Dr. DeBakey had told him it was time for nature to take its course, but also told him that the doctors had "to do what we need to do." Members of Dr. DeBakey's medical team said they interpreted the statements differently. Some thought he meant that they should do watchful waiting, acting only if conditions warranted; others thought it meant he wanted to die.

The question was whether the operation would counter Dr. DeBakey's wishes expressed in his signed "do not resuscitate" order. Some said that everything Dr. DeBakey did was for his family. And the family wanted the operation.

After the committee members had met for an hour, Mrs. DeBakey could stand it no longer. She charged into the room.

"My husband's going to die before we even get a chance to do anything—let's get to work," she said she told them.

The discussion ended. The majority ruled in a consensus without a formal vote. No minutes were kept, the doctors said.

"Boy, when that meeting was over, it was single focus—the best operation, the best post-operative care, the best recovery we could give him," Dr. Pool said.

AS THE ETHICS COMMITTEE MEETING ended about 11 p.m. on Feb. 9, the doctors rushed to start Dr. DeBakey's anesthesia.

The operation was to last seven hours.

For part of that time, Dr. DeBakey's body was cooled to protect his brain and other organs. His heart was stilled while a heart-lung bypass machine pumped oxygen-rich blood through his body. The surgeons replaced the damaged portion of Dr. DeBakey's aorta with a six- to eight-inch graft made of Dacron, similar to material used in shirts. The graft was the type that Dr. DeBakey devised in the 1950s.

Afterward, Dr. DeBakey was taken to an intensive care unit.

Some doctors were waiting for Dr. DeBakey to die during the operation or soon thereafter, Dr. Noon said. "But he just got better."

As feared, however, his recovery was stormy.

Surgeons had to cut separate holes into the trachea in his neck and stomach to help him breathe and eat. He needed dialysis because of kidney failure. He was on a mechanical ventilator for about six weeks because he was too weak to breathe on his own. He developed infections. His blood pressure often fell too low when aides lifted him to a sitting position. Muscle weakness left him unable to stand.

For a month, Dr. DeBakey was in the windowless intensive care unit, sometimes delirious, sometimes unresponsive, depending in part on his medications. The doctors were concerned that he had suffered severe, permanent brain damage. To allow him to tell day from night and lift his spirits, the hospital converted a private suite into an intensive care unit.

Some help came from unexpected places. On Sunday, April 2, Dr. William W. Lunn, the team's lung specialist, took his oldest daughter, Elizabeth, 8, with him when he made rounds at the hospital and told her that a patient was feeling blue. While waiting, Elizabeth drew a cheery picture of a rainbow, butterflies, trees and grass and asked her father to give it to the patient. He did.

"You should have seen Dr. DeBakey's eyes brighten," Dr. Lunn said. Dr. DeBakey asked to see Elizabeth, held her hand and thanked her.

"At that point, I knew he was going to be O.K.," Dr. Lunn said.

Dr. DeBakey was discharged on May 16. But on June 2, he was back in the hospital.

"He actually scared us because his blood pressure and heart rate were too high, he was gasping for breath" and he had fluid in his lungs, Dr. Lunn said.

But once the blood pressure was controlled with medicine, Dr. DeBakey began to recover well.

AT TIMES, Dr. DeBakey says he played possum with the medical team, pretending to be asleep when he was listening to conversations.

On Aug. 21, when Dr. Loebe asked Dr. DeBakey to wake up, and he did not, Dr. Loebe announced that he had found an old roller pump that Dr. DeBakey devised in the 1930s to transfuse blood. Dr. DeBakey immediately opened his eyes. Then he gave the doctors a short lecture about how he had improved it over existing pumps.

As he recovered and Dr. DeBakey learned what had happened, he told his doctors he was happy they had operated on him. The doctors say they were relieved because they had feared he regretted their decision.

"If they hadn't done it, I'd be dead," he said.

The doctors and family had rolled the dice and won.

Dr. DeBakey does not remember signing an order saying not to resuscitate him and now thinks the doctors did the right thing. Doctors, he said, should be able to make decisions in such cases, without committees.

Throughout, Dr. DeBakey's mental recovery was far ahead of his physical response.

When Dr. DeBakey first became aware of his post-operative condition, he said he "felt limp as a rag" and feared he was a quadriplegic. Kenneth Miller and other physical therapists have helped Dr. DeBakey strengthen his withered muscles.

"There were times where he needed a good bit of encouragement to participate," Mr. Miller said. "But once he saw the progress, he was fully committed to what we were doing."

Now he walks increasingly long distances without support. But his main means of locomotion is a motorized scooter. He races it around corridors, sometimes trailed by quick-stepping doctors of all ages.

Dr. DeBakey said he hoped to regain the stamina to resume traveling, though not at his former pace.

Dr. William L. Winters Jr., a cardiologist on Dr. DeBakey's team, said: "I am impressed with what the body and mind can do when they work together. He absolutely has the desire to get back to where he was before. I think he'll come close."

Already, Dr. DeBakey is back working nearly a full day.

"I feel very good," he said Friday. "I'm getting back into the swing of things."

Elizabeth Kolbert

Butterfly Lessons

FROM *THE NEW YORKER*

Butterflies migrating to locations further north, mosquitoes laying their eggs later in the year, mountain-dwelling toads going extinct. Following researchers covering these species, Elizabeth Kolbert finds that global warming is behind all of these troubling changes.

POLYGONIA C-ALBUM, generally known as the comma butterfly, spends most of its life pretending to be something else. In its larval, or caterpillar, stage, it has a chalky stripe down its back which makes it look uncannily like a bird dropping. As an adult, with wings folded, it is practically indistinguishable from a dead leaf. The comma gets its name from a tiny white mark shaped like the letter "c" on its underside. Even this is thought to be part of its camouflage—an ersatz rip of the sort that leaves get when they are particularly old and tatty.

The comma is a European butterfly—its American cousins are the hop merchant and the question mark—and it can be found in France, where it is known as *le Robert-le-Diable*; Germany, where it is called *der C-Falter*; and the Netherlands, where it is *gehakkelde aurelia*. The comma reaches the northern edge of its distribution in Britain. This is unremarkable—many European butterflies come to the end of their range in England—but from a scientific standpoint fortunate.

The English have been watching and collecting butterflies for centuries—some of the specimens in the British Natural History Museum date back to the seventeen-hundreds—and in the Victorian era passion for the hobby was such that even many small towns supported their own entomological societies. In the nineteen-sixties, Britain's Biological Records Centre decided to marshal this enthusiasm for a project called the Lepidoptera Distribution Maps Scheme, whose aim was to chart precisely where each of the country's fifty-nine native species could—and could not—be found. More than two thousand butterfly enthusiasts participated, and in 1984 the results were collated into a hundred-and-fifty-eight-page atlas. Every species got its own map, with black dots showing where it had been sighted. On the map for *Polygonia c-album*, the comma's range was shown to extend from the south coast of England up to Liverpool in the west and Norfolk in the east. Almost immediately, the map became out of date; in the years that followed, hobbyists kept finding the comma in new areas. By the late nineteen-nineties, the butterfly was frequently being sighted in the north of England, near Durham. By now, it is established in southern Scotland, and has been sighted as far up as the Highlands. The rate of the comma's expansion—some fifty miles per decade—was described by the authors of the most recent butterfly atlas as "remarkable."

Chris Thomas is a biologist at the University of York who studies lepidoptera. He is tall and rangy, with an Ethan Hawke-style goatee and an amiably harried manner. The day I met him, he had just returned from looking for butterflies in Wales, and the first thing he said to me when I got into his car was please not to mind the smell of wet socks. A few years ago, Thomas, his wife, their two sets of twins, an Irish wolfhound, a pony, some rabbits, two cats, and several chickens moved into an old farmhouse in the village of Wistow, in the Vale of York. The University of York has an array of thermostatic chambers where commas are raised under temperature-controlled conditions, fed carefully monitored diets, and measured on a near-constant basis, but, in the spirit of British amateurism, Thomas decided to turn his own back yard into a field lab. He scattered wildflower seeds he had collected from nearby meadows and ditches, planted nearly seven hundred trees, and waited for the butterflies to show up. When I visited the place in midsummer, the wildflowers were in bloom and the grass was so high that many of the tiny trees looked lost, like kids in search of their parents. The Vale of York is almost completely flat—during the last ice age, it formed the bottom of a giant lake—and from the yard Thomas pointed out the spires of Selby Abbey, built nearly a thousand years ago, and also the cooling towers of the Drax power plant, Britain's largest, some ten miles away. It was cloudy, and since butterflies don't generally fly when it's gray, we went inside.

Butterflies, Thomas explained after putting the kettle on for tea, can be divided into two groups. The so-called "specialists" require specific—in some cases unique—conditions. Specialists include the chalkhill blue (*Polyommatus coridon*), a large turquoise butterfly that feeds exclusively on horseshoe vetch, and the purple emperor (*Apatura iris*), which flies in the treetops of well-wooded areas in southern England. The "generalists" are less picky. Among Britain's generalists, there are, in addition to the comma, ten species that are widespread in the southern part of the country and reach the edge of their range somewhere in the nation's midsection. "Every single one has moved northward since 1982," Thomas told me. A few years ago, with lepidopterists from, among other places, the United States, Sweden, France, and Estonia, Thomas conducted a survey of all the studies that had been done on generalists that reach the northern limits of their ranges in Europe. The survey looked at thirty-five species in all. Of these, the scientists found, twenty-two had shifted their range northward in recent decades; only one had shifted south.

After a while, the sun emerged, and we went back outside. Thomas's wolfhound, Rex, a dog the size of a small horse, trailed behind us, panting heavily. Within about five minutes, Thomas had identified a meadow brown (*Maniola jurtina*), a small tortoiseshell (*Aglais urticae*), and a green-veined white (*Pieris napi*), all species that have been flitting around Yorkshire since butterfly recordkeeping began. Thomas also spotted a gatekeeper (*Pyronia tithonus*) and a small skipper (*Thymelicus sylvestris*), which until recently had been confined to a region south of where we were standing. "So far, two out of the five species of butterflies that we've seen are northward invaders," he told me. "Sometime within the last thirty years they have spread into this area." A few minutes later, he pointed out another invader sunning itself in the grass—a *Polygonia c-album*. With its wings closed, the comma was a dull, dead-leaf brown, but with them open it was a brilliant orange.

That life on earth changes with the climate has been assumed to be the case for a long time—indeed, for very nearly as long as the climate has been known to be capable of changing. In 1840, Louis Agassiz published *Études sur les Glaciers*, the work in which he laid out his theory of the ice ages. By 1859, Charles Darwin had incorporated Agassiz's theory into his own theory of evolution. Toward the end of *On the Origin of Species*, in a chapter titled "Geographical Distribution," Darwin describes the vast migrations that he supposes the advance and retreat of the glaciers must have necessitated:

As the cold came on, and as each more southern zone became fitted for arctic beings and ill-fitted for their former more temperate inhabitants, the latter would be supplanted and arctic productions would take their places. The inhabitants of the more temperate regions would at the same time travel southward. . . . As the warmth returned, the arctic forms would retreat northward, closely followed up in their retreat by the productions of the more temperate regions. And as the snow melted from the bases of the mountains, the arctic forms would seize on the cleared and thawed ground, always ascending higher and higher, as the warmth increased, whilst their brethren were pursuing their northern journey.

The last of the great ice sheets retreated some ten thousand years ago, at the start of the Holocene. At that point, the concentration of carbon dioxide in the atmosphere stood at two hundred and sixty parts per million. Give or take twenty parts per million, it remained at that level through the invention of agriculture, the founding of the first cities, the building of the pyramids, and the discovery of the New World. When, in the early eighteen-hundreds, coal-burning began to drive up CO_2 levels, they rose at first gradually—it took more than a century to reach three hundred parts per million—and later, following the Second World War, much more rapidly. By 1965, CO_2 concentrations had reached three hundred and twenty parts per million; by 1985, three hundred and forty-six parts per million; and, by 2005, three hundred and seventy-eight parts per million. If current trends continue, it will reach five hundred parts per million (nearly double pre-industrial levels) by the middle of this century, and could reach as much as seven hundred and fifty parts per million (nearly triple pre-industrial levels) by 2100. The equilibrium warming associated with doubled CO_2 is estimated to be between three and a half and seven degrees, and with tripled CO_2 between six and eleven degrees. A global temperature rise of just three degrees would render the earth hotter than it has been at any point in the past two million years.

This vast geophysical experiment is a biological one as well. Darwin never imagined that the effects of climate change could be observed in a human lifetime, yet, almost anywhere you go in the world today, it is possible to observe changes comparable to the northern expansion of the comma. A recent study of common frogs living near Ithaca, New York, for example, found that four out of six species were calling, which is to say mating, at least ten days earlier than at the start of the nineteen-hundreds, while at the Arnold Arboretum, in Boston, the peak blooming date for spring-flowering shrubs has advanced, on average, by eight days. In Costa Rica, birds like the keel-billed toucan (*Ramphastos sulfuratus*), once confined to the lowlands and foothills, have started to nest on mountain slopes; in the Alps, plants like purple saxifrage (*Saxifraga oppositifolia*) and

Austrian draba (*Draba fladnizensis*) have been creeping up toward the summits; and in the Sierra Nevada mountains of California the average Edith's checker-spot butterfly (*Euphydryas editha*) is now found at an elevation three hundred feet higher than it was a hundred years ago. To what extent life on earth will be transformed by the warming expected in the coming years is, at this point, still a matter of speculation. Clearly, though, the process has begun.

THE BRADSHAW-HOLZAPFEL LAB occupies a corner on the third floor of Pacific Hall, a peculiarly unlovely building on the campus of the University of Oregon, in Eugene. At one end of the lab is a large room stacked with glassware, and at the other end is a trio of offices. In between are several workrooms that look, from the outside, like walk-in refrigerators. Taped to the door of one of them is a handwritten sign: "Warning—If you enter this room mosquitoes will suck your blood out through your eyes!"

William Bradshaw and Christina Holzapfel, who run the lab and share one of the offices, are evolutionary geneticists. They met as graduate students at the University of Michigan, and have been married for thirty-four years. Bradshaw is a tall man with thinning gray hair and a gravelly voice. His desk is covered in a mess of papers, books, and journals, and when visitors come to the lab he likes to show them his collection of curiosities, which includes a desiccated octopus. Holzapfel is short, with blond hair and bright-blue eyes. Her desk is perfectly neat.

Bradshaw and Holzapfel have shared an interest in mosquitoes for as long as they've been interested in each other. In the early years of their lab, which they set up in 1971, they raised several different species, some of which, in order to reproduce, required what is delicately referred to as a "blood meal." This, in turn, demanded a live animal able to provide such a meal. For a time, this requirement was met by rats anesthetized with phenobarbital, but, as rules about experimenting with animals grew more stringent, Bradshaw and

Holzapfel found themselves forced to decide whether it was more humane to keep anesthetizing the same rat over and over or to use a new rat and let the old one wake up to find itself covered with bites. Eventually, they decided to stick to a single species, *Wyeomyia smithii*, which needs no blood in order to reproduce. At any given moment, the Bradshaw-Holzapfel lab houses upward of a hundred thousand *Wyeomyia smithii* in various stages of development.

Wyeomyia smithii is a small and rather ineffectual bug. ("Wimpy" is how Bradshaw characterizes it.) Its eggs are practically indistinguishable from specks of dust; its larvae appear as minuscule white wrigglers. As an adult, it is about a quarter of an inch long and in flight looks like a tiny black blur. Only when you examine a *Wyeomyia smithii* very closely, under a magnifying glass, can you see that its abdomen is actually silver and that its two hind legs are bent gracefully above its head, like a trapeze artist's.

Wyeomyia smithii completes virtually its entire life cycle—from egg to larva to pupa to adult—inside a single plant, *Sarracenia purpurea*, or, as it is more commonly known, the purple pitcher plant. The purple pitcher plant, which grows in swamps and peat bogs from Florida to northern Canada, has frilly, cornucopia-shaped leaves that sprout directly out of the ground and then fill with water. In the spring, female *Wyeomyia smithii* lay their eggs one at a time, carefully depositing them in different pitcher plants. When ants and flies and, occasionally, small frogs drown in the leaves of the pitcher plant, their remains provide nutrients not only for the plant—*Sarracenia purpurea* is carnivorous—but also for developing mosquito larvae. (*Sarracenia purpurea* does not digest its own food; it leaves this task to bacteria, which don't attack the mosquitoes.) Once the young mature into adults, they repeat the whole process, and, if conditions are right, the cycle can be completed four or five times in a single summer. Come fall, the adult mosquitoes die off, but the larvae live on through the winter in a state of suspended animation known as diapause—the insect version of hibernation.

The exact timing of diapause is critical to the survival of *Wyeomyia smithii* and also to Bradshaw and Holzapfel's research. When the larvae perceive that day length has dropped below a certain threshold, they stop growing and molting; when they perceive that it has lengthened sufficiently, they take up where they left off.

This light threshold, which is known as the critical photoperiod, varies from bog to bog. At the southern end of the mosquitoes' range, near the Gulf of Mexico, conditions remain favorable for breeding well into fall. A typical *Wyeomyia smithii* from Florida or Alabama will, consequently, not go dormant until day length has shrunk to about twelve hours, which at that latitude corresponds to early November. At the far northern edge of the range, meanwhile, winter arrives much earlier, and an average mosquito from Manitoba will go into dormancy in late July, as soon as day length drops below sixteen and a half hours. Interpreting light cues is a genetically controlled and highly heritable trait: *Wyeomyia smithii* are programmed to respond to day length the same way their parents did, even if they find themselves living under very different conditions. (One of the walk-in-freezer-like rooms in the Bradshaw-Holzapfel lab contains locker-size storage units, each equipped with a timer and a fluorescent bulb, where mosquito larvae can be raised under any imaginable schedule of light and dark.) In the mid-nineteen-seventies, Bradshaw and Holzapfel demonstrated that *Wyeomyia smithii* living at different elevations also obey different light cues—high-altitude mosquitoes behave as if they were born farther north—a discovery that today might seem relatively unremarkable but at the time was sufficiently noteworthy to make the cover of *Nature*.

About five years ago, Bradshaw and Holzapfel began to wonder about how *Wyeomyia smithii* might be affected by global warming. They knew that the species had expanded northward after the end of the last glaciation, and that, at some point in the intervening millennia, the critical photoperiods of northern and southern populations had diverged. If the climatic conditions for *Wyeomyia smithii* were changing once again, then perhaps this would show up in the timing of diapause. The first thing the couple did was go back to look at their old records, to see if the data contained any information that they hadn't noticed before.

"There it was," Holzapfel told me. "Just hitting you right in the eye."

WHEN AN ANIMAL CHANGES ITS ROUTINE by, say, laying its eggs earlier or going into hibernation later, there are a number of possible explanations. One is that the change reflects an innate flexibility; as conditions vary, the animal is able to adjust its behavior in response. Biologists call such flexibility "phenotypic plasticity," and it is key to the survival of most species. Another possibility is that the shift represents something deeper and more permanent—an actual rearrangement of the organism's genetic structure.

In the years since Bradshaw and Holzapfel established their lab, they have collected mosquito larvae from all over the eastern United States and much of Canada. They used to do the collecting themselves, driving across the country in a van equipped with a makeshift bed for their daughter and a miniature lab for sorting, labelling, and storing the thousands of specimens they would gather. Nowadays, they more often send out their graduate students, who, instead of driving, are likely to fly. (Getting through airport security with a backpack full of mosquito larvae is a process that, the students have learned, can take half a day.)

Every subpopulation exhibits a range of light responses; Bradshaw and Holzapfel define critical photoperiod as the point at which fifty per cent of the mosquitoes in a sample have switched from active development to diapause. Each time they collect a new batch of insects, they put the larvae in petri dishes and place the dishes in the controlled-environment light boxes, which they call Mosquito Hiltons. Then they test the larvae for their critical photoperiod, and record the results.

When Bradshaw and Holzapfel went back to their files, they looked for populations that they had tested repeatedly. One of these was from a wetland called Horse Cove, in Macon County, North Carolina. In 1972, when they had collected mosquitoes from Horse Cove for the first time, their files showed, the larvae's critical photoperiod was fourteen hours and twenty-one minutes. They collected a second batch of mosquitoes from the same spot in 1996. By that point, the insects' critical photoperiod had dropped to thirteen hours and fifty-three minutes. All told, Bradshaw and Holzapfel found comparative data in their files on ten different subpopulations—two in Florida, three in North Carolina, two in New Jersey, and one each in Alabama, Maine, and Ontario. In every case, the critical photoperiod had declined over time. Their data also showed that the farther north you went the stronger the effect; a regression analysis revealed that the critical photoperiod of mosquitoes living at fifty degrees north latitude had declined by more than thirty-five minutes, corresponding to a delay in diapause of nine days.

In a different mosquito, this shift could be an instance of the kind of plasticity that allows organisms to cope with varying conditions. But in *Wyeomyia smithii* there is almost no flexibility when it comes to timing the onset of diapause. Warm or cold, all the insect can do is read light. Bradshaw and Holzapfel knew, therefore, that the change they were seeing must be genetic. As the climate had warmed, those mosquitoes which had remained active until later in the fall had enjoyed a selective advantage, presumably because they had been able to store a few more days' worth of resources for the winter, and they had passed this advantage on to their offspring, and so on. In December, 2001, Bradshaw and Holzapfel published their findings in the *Proceedings of the National Academy of Sciences.* By doing so, they became the first researchers to demonstrate that global warming had begun to drive evolution.

THE MONTEVERDE CLOUD FOREST sits astride the Cordillera de Tilarán, or Tilarán Mountains, in northwestern Costa Rica. The rugged terrain, in combination with the trade winds that blow off the Caribbean, makes the region unusually diverse; in an area of less than two hundred and fifty square miles, there are seven different "life zones," each with its own distinctive type of vegetation. The cloud forest is surrounded on all sides by land, yet, ecologically speaking, it is an island and, like many islands, displays a high degree of endemism, or biological specificity. Fully ten per cent of Monteverdean flora, for example, are believed to be unique to the Cordillera de Tilarán.

The most famous of Monteverde's endemic species is a small toad. Known colloquially as the golden toad, it was officially discovered by a biologist from the University of Southern California named Jay Savage. Savage had heard of the toad from a local resident who lived in a Quaker community at the edge of the forest; still, when he came across it for the first time, on May 14, 1964, at the top of a high mountain ridge, his reaction, he would later recall, was one of "disbelief." Most toads are dull brown, grayish green, or olive; this one was a flaming shade of tangerine. Savage named the new species *Bufo periglenes*, from a Greek word meaning "bright," and titled his paper on the discovery "An Extraordinary New Toad (*Bufo*) from Costa Rica."

Since the golden toad spent its life underground, emerging only in order to reproduce, most of what was subsequently learned about it had to do with sex. The toad turned out to be an "explosive breeder." Instead of staking out and defending territory, males simply rushed the first available female and fought for the chance to mount her. ("Amplexus" is the term of art for an amphibian embrace.) Generally, males outnumbered females, in some years by as much as ten to one, a situation that often led bachelors to attack amplectant pairs and form what Savage once described as "writhing masses of toad balls." The eggs of the golden toad, black-and-tan spheres, were deposited in small pools—puddles, really—often no more than an inch deep. Tadpoles emerged in a matter of days, but required another month for metamorphosis. During this period, they were highly dependent on the weather: too much rain and they would be washed down the steep hillsides, too little and their puddles would dry up. Golden toads were never found more than a few miles from the site where Savage originally spotted them, always at the top of a mountain ridge, and always at an altitude of between forty-nine hundred and fifty-six hundred feet.

In the spring of 1987, an American biologist who had come to the cloud forest specifically to study the amphibians there counted fifteen hundred golden toads in temporary breeding pools. That spring was unusually warm and dry, and most of the pools evaporated before the tadpoles in them had had time to mature. The following year, only one male was seen at what previously had been the major breeding site. Seven males and two females were seen at a second site a few miles away. The year after that, a search of all spots where the toad had earlier been sighted yielded a solitary male. No golden toad has been seen since, and it is widely assumed that after living its colorful, if secretive, existence for hundreds of thousands of years *Bufo periglenes* is now extinct.

In April, 1999, J. Alan Pounds, who heads the Golden Toad Laboratory for Conservation, in the Monteverde Preserve, published a paper in *Nature* on the golden toad's demise. In it, he linked the toad's extinction, as well as the decline of several other amphibian species, to a shift in rainfall patterns. In recent years, there has been a significant increase in the number of days with no measurable precipitation in Monteverde, a change that, in turn, is consonant with an increase in the elevation of the cloud cover. In a separate article in the same issue of *Nature*, a group of scientists from the United States and Japan reported on their efforts to model the future of cloud forests. They predicted that as global CO_2 levels continued to rise, the height of the cloud cover in Monteverde and other tropical cloud forests would continue to climb. This, they speculated, would force additional high-altitude species "out of existence."

CLIMATE CHANGE—even violent climate change—is itself, of course, part of the natural order. In the past two million years, great ice sheets have advanced over the Northern Hemisphere and retreated again at least twenty times. In addition, there have also been dozens of abrupt climate shifts, like the Younger Dryas, which occurred some twelve thousand eight hundred years ago. (The event is named after a small Arctic plant—*Dryas octopetala*—that suddenly reappeared in Scandinavia.) At that point, the earth, which had been warming rapidly, cooled back down into ice-age conditions. It remained frigid for twelve centuries and then warmed, even more abruptly; in Greenland, ice-core records show, average annual temperatures climbed by nearly twenty degrees in a single decade.

Thompson Webb III is a paleoecologist who teaches at Brown University. He studies pollen grains and fern spores, in an effort to reconstruct the plant life of previous eras. In the mid-seventies, Webb began to assemble a database of pollen records from lakes all across North America. (When a grain of pollen falls on the ground, it usually oxidizes and disappears; if it is blown onto a body of water, however, it can sink to the bottom and be preserved in the sediment for millennia.) The project took nearly twenty years to complete, and, when it was finally done, it showed how, as the climate of the continent had changed, life had rearranged itself.

A few months ago, I went to visit Webb in Providence. He has an office in the university's geochemistry building, and also a lab, where, on this particular day, one of his research assistants was examining charcoal particles from an ancient forest fire. Webb took some slides from a cabinet and slipped one under the lens of a microscope. Most pollen grains are between twenty and seventy microns in diameter; to be identified, they must be magnified four hundred times. Peering through the eyepiece, I saw a tiny sphere, pocked like a golf ball. Webb told me that what I was looking at was a grain of birch pollen. He replaced the slide, and a second tiny golf ball swam into focus. It was beech pollen, Webb explained, and could be distinguished by a set of three minute grooves. "You see, they're really very different," he said of the two grains.

After a while, we went down the hall to Webb's office. On his computer he called up a program named Pollen Viewer 3.2, and a map of North America circa 19000 B.C. appeared on the screen. Around that time, the ice sheets of the last glaciation reached their maximum extent; the map showed the Laurentide ice sheet covering all of Canada as well as most of New England and the upper Midwest. Because so much water was tied up in the ice, sea levels were some three hundred feet lower than they are now. On the map, Florida appeared as a stubby protuberance, nearly twice as wide as it is today. Webb clicked on "Play." Time began to move forward in thousand-year increments. The ice sheet shrank. A huge lake, known as Lake Agassiz, formed in central Canada, and, a few thousand years later, drained. The Great Lakes emerged, and then widened. Around eight thousand years ago, open water finally appeared in Hudson Bay. The bay began to contract as the land around it rebounded from the weight of the ice sheet.

Webb clicked on a pull-down menu that listed the Latin names of dozens of trees and shrubs. He chose *Pinus* (pine) and again hit "Play." Dark-green splotches began to move around the continent. Twenty-one thousand years ago, the program showed, pine forests covered the entire Eastern Seaboard south of the ice sheet. Ten thousand years later, pines were concentrated around the Great Lakes, and today pine predominates in the southeastern United States and in western Canada. Webb clicked on *Quercus* (oak), and a similar process began, only *Quercus* moved in a very different pattern from *Pinus*. More clicks for *Fagus* (beech), *Betula* (birch), and *Picea* (spruce). As the earth warmed and the continent emerged from the ice age, each of the tree species migrated, but no two moved in exactly the same way.

"The trick you've got to remember is that climate is multivariate," Webb explained. "The plant species are having to respond both to temperature changes and to moisture changes and to changes in seasonality. It makes a big difference if you have a drier winter versus a drier summer, because some species are more attuned to spring and others to fall. Any current community has a certain mixture, and, if you start changing the climate, you're changing the temperature, but you're also changing moisture or the timing of the moisture or the amount of snow and, bingo, species are not going to move together. They can't."

Webb pointed out that the warming predicted for the next century is on the same scale as the temperature difference between the last glaciation and today. "You know that's going to give us a very different landscape," he said. I asked what he thought this landscape would look like. He said he didn't know—his central finding, from more than thirty years of research, is that, as the climate changes, species often move in surprising ways. In the short term, which is to say in the remainder of his own life, Webb said that he expected mostly to see disruption.

"We have this strange sense of the evolutionary hierarchy, that the microorganisms, because they came first, are the most primitive," he told me. "And yet you could argue that this will just give a lot of advantage to the microorganisms of the world, because of their ability to evolve more quickly. To the extent the climate is putting organisms as well as ecosystems under stress, it's opening the opportunities for invasive species on the one hand and disease on the other. I guess I start thinking: Think death."

ANY SPECIES THAT IS AROUND TODAY, including our own, has already survived catastrophic climate change. The fact that a species has survived such a change, or even many such changes, is no guarantee, however, that it will survive the next one. Consider, for example, the outsized megafauna—seven-hundred-and-fifty-pound sabre-toothed cats, elephantine sloths, and fifteen-foot-tall mastodons—that once dominated the North American landscape. These mega-fauna lived through several glacial cycles, but then something changed, and they nearly all died out at the same time, at the beginning of the Holocene.

Over the past two million years, even as the temperature of the earth has swung wildly, it has always remained within certain limits: the planet has often been colder than today, but rarely warmer, and then only slightly. If the earth continues to warm at the current rate, then by the end of this century temperatures will push beyond the "envelope" of natural climate variability.

Meanwhile, thanks to us, the world today is a very different—and in many ways diminished—place. International trade has introduced exotic pests and competitors; ozone depletion has increased exposure to ultraviolet radiation; and many species have already been very nearly wiped out, or wiped out altogether, by overhunting and over-harvesting. Perhaps most significantly, human activity, in the form of farms and cities and subdivisions and mines and logging operations and parking lots, has steadily reduced the amount of available habitat. G. Russell Coope is a visiting professor in the geography department at the University of London and one of the world's leading authorities on ancient beetles. He has shown that, under the pressure of climate change, insects have migrated tremendous distances; for example, *Tachinus caelatus*, a small, dullish-brown beetle common in England during the cold periods of the Pleistocene, today can be found only some five thousand miles away, in the mountains west of Ulan Bator, in Mongolia. But Coope questions whether such long-distance migrations are practical in a fragmented landscape like today's. Many organisms now live in the functional equivalent of "oceanic islands or remote mountain tops," he has written. "Certainly, our knowledge of their past response may be of little value in predicting any future reactions to climate change, since we have imposed totally new restrictions on their mobility; we have inconveniently moved the goal posts and set up a ball game with totally new rules."

A few years ago, nineteen biologists from around the world set out to give, in their words, a "first pass" estimate of the extinction risk posed by global warming. They assembled data on eleven hundred species of plants and animals from sample regions covering roughly a fifth of the earth's surface. Then they established the species' current ranges, based on climate variables such as temperature and rainfall. Finally, they calculated how much of the species' "climate envelope" would be left under different warming scenarios. The results of this effort were published in *Nature* in 2004. Using a mid-range projection of temperature rise, the biologists concluded that, if the species in the sample regions could be assumed to be highly mobile, then fully fifteen per cent of them would be "committed to extinction" by the middle of this century, and, if they proved to be basically stationary, an extraordinary thirty-seven per cent of them would be.

THE MOUNTAIN RINGLET (*Erebia epiphron*) is a dun-colored butterfly with orange-and-black spots that curl along the edges of its rounded wings. It overwinters as a larva, and as an adult has an extremely brief life span—perhaps as short as one or two days. A montane, or mountain, species, it is found only at elevations above a thousand feet in the Scottish Highlands and, farther south, in Britain's Lake District, only above fifteen hundred feet.

Together with a colleague from the University of York, Chris Thomas has for the past few years been monitoring the mountain ringlet, along with three other species of butterfly—the Scotch Argus (*Erebia aethiops*), the large heath (*Coenonympha tullia*), and the northern brown Argus (*Aricia artaxerxes*)—whose ranges are similarly confined to northern England and Scotland. In the summer of 2004, researchers for the project visited more than four hundred sites where these "specialist" species had been sighted in the past, and last summer they repeated the process. Documenting a species' contraction is more difficult than documenting its expansion—is it really gone, or did someone just miss it?—but preliminary evidence suggests that the butterflies are already disappearing from lower-elevation and more southerly sites. When I went to visit Thomas, he was getting ready to take his family to Scotland on vacation and was planning to recheck some of the sites. "It's a bit of a busman's holiday," he confessed.

As we were wandering around his yard in search of commas, I asked Thomas, who was the lead author on the extinction study, how he felt about the changes he was seeing. He told me that he found the opportunities for study presented by climate change to be exciting.

"Ecology for a very long time has been trying to explain why species have the distribution that they do, why a species can survive here and not over there, why some species have small distributions and others have broad ones," he said. "And the problem that we have always had is that distributions have been rather static. We couldn't actually see the process of range boundaries changing taking place, or see what was driving those changes. Once everything starts moving, we can begin to understand: is it a climatic determinant or is it mainly other things, like interactions with other species? And, of course, if you think of the history of the last million years, we now have the opportunity to try and understand how things might have responded in the past. It's extremely interesting—the prospect of everything changing its distribution, and new mixtures of species from around the world starting to form and produce new biological communities. Extremely interesting from a purely academic point of view.

"On the other hand, given our conclusions about possible extinctions, it is, to me personally, a serious concern," he went on. "If we are in the situation where a quarter of the terrestrial species might be at risk of extinction from climate change—people often use the phrase of being like canaries—if we've changed our biological system to such an extent, then we do have to get worried about whether the services that are provided by natural ecosystems are going to continue. Ultimately, all of the crops we grow are biological species; all the diseases we have are biological species; all the disease vectors are biological species. If there is this overwhelming evidence that species are changing their distributions, we're going to have to expect exactly the same for crops and pests and diseases. Part of it simply is we've got one planet, and we are heading it in a direction in which, quite fundamentally, we don't know what the consequences are going to be."

WILLIAM J. BROAD

In Ancient Fossils, Seeds of a New Debate on Warming

FROM THE *NEW YORK TIMES*

As consensus grows among scientists that carbon dioxide and other man-made greenhouse gases are responsible for global warming, new research, stretching deep into the earth's past, may complicate the picture. As William J. Broad reports, both proponents of the greenhouse theory of global warming and skeptics are finding support for their points of view in evidence of high carbon-dioxide concentrations millions of years ago.

I N RECENT YEARS, scientists have made sizable gains in what was once considered an impossible art —reconstructing the history of Earth's atmosphere back into the dim past. They can now peer across more than a half billion years.

The scientists have learned about the changing makeup of the vanished gases by teasing subtle clues from fossilized soils, plants and sea creatures. They have also gained insights from computer models that predict how phenomena like eroding rocks and erupting volcanoes have altered the planet's evolving air. "It's getting a lot more attention," Michael C. MacCracken, chief scientist of the Climate Institute, a research group in Washington, said of the growing field.

For the first time, the Intergovernmental Panel on Climate Change, a United Nations group that analyzes global warming, plans to include a chapter on the reconstructions in its latest report, due early next year.

The discoveries have stirred a little-known dispute that, if resolved, could have major implications. At issue is whether the findings back or undermine the prevailing view on global warming. One side foresees a looming crisis of planetary heating; the other, temperature increases that would be more nuisance than catastrophe.

Perhaps surprisingly, both hail from the same camp: scientists who study the big picture of Earth's past, including geologists and paleoclimatologists.

Most public discussions of global warming concentrate on evidence from the last few hundred or, at most, few thousand years. And some climate scientists remain unconvinced that data from the deep past are solid enough to be relevant to the debates.

But the experts who peer back millions of years, though they may debate what their work means, do agree on the relevance of their findings. They also agree that the eon known as the Phanerozoic, a lengthy span from the present to 550 million years ago, the dawn of complex life, typically bore concentrations of carbon dioxide that were up to 18 times the levels present in the short reign of *Homo sapiens.*

The carbon dioxide, the scientists agree, came from volcanoes and other natural sources, as on Mars and Venus. The levels have generally dropped over the ages, as the carbon became a building block of many rock formations and all living things.

Moreover, the opponents tend to agree on why the early Earth's high carbon dioxide levels failed to roast the planet. First, the Sun was dimmer in its youth. Second, as the gas concentrations increase, its heat trapping capacity slows and reaches a plateau.

Where the specialists clash is on what the evidence means for the idea that industrial civilization and the burning of fossil fuels are the main culprits in climate change. The two sides agree that carbon dioxide can block solar energy that would otherwise radiate back into space, an effect known as greenhouse warming. But they differ sharply on its strength.

Some argue that CO_2 fluctuations over the Phanerozoic follow climate trends fairly well, supporting a causal relationship between high gas levels and high temperatures. "The geologic record over the past 550 million years indicates a good correlation," said Robert A. Berner, a Yale geologist and pioneer of paleoclimate analysis. "There are other factors at work here. But in general, global warming is due to CO_2. It was in the past and is now."

Other experts say that is an oversimplification of a complex picture of natural variation. The fluctuations in the gas levels, they say, often fall out of step with the planet's hot and cold cycles, undermining the claimed supremacy of carbon dioxide.

"It's too simplistic to say low CO_2 was the only cause of the glacial periods" on time scales of millions of years, said Robert Giegengack, a geologist at the University of Pennsylvania who studies past atmospheres. "The record violates that one-to-one correspondence."

He and other doubters say the planet is clearly warming today, as it has repeatedly done, but insist that no one knows exactly why. Other possible causes, they say, include changes in sea currents, Sun cycles and cosmic rays that bombard the planet.

"More and more data," Jan Veizer, an expert on Phanerozoic climates at the University of Ottawa, said, "point to the Sun and stars as the dominant driver."

Highlighting the gap, the two sides clash on how much the Earth would warm today if carbon dioxide concentrations double from preindustrial levels, as scientists expect. Many climatologists see an increase of as much as 8 degrees Fahrenheit. The skeptics, drawing on Phanerozoic data, tend to see far less, perhaps 2 or 3 degrees.

In the Phanerozoic (the term is Greek for visible life), complex organisms arose. If its countless ages were compressed into a single year, fish would have appeared in January, land animals in March, dinosaurs in June, monkeys in December and humans late on New Year's Eve.

The Phanerozoic dispute, fought mainly in scholarly journals and scientific meetings, has occurred in isolation from the public debate on global warming. Al Gore in *An Inconvenient Truth* makes no mention of it.

Some mainstream scientists familiar with the Phanerozoic evidence call it too sketchy for public consumption and government policy, if not expert deliberations.

"In my view, the uncertainties are too great to draw any conclusions right now," Michael Oppenheimer, a professor of geosciences and international affairs at Princeton, said. "It could be that when the dust settles some insight will emerge that will be germane to the current problem—how do we keep the climate from spinning out of control?"

Skeptics say CO_2 crusaders simply find the Phanerozoic data embarrassing and irreconcilable with public alarms. "People come to me and say, 'Stop talking like this, you're hurting the cause,'" said Dr. Giegengack of Penn.

Robert A. Rohde, a graduate student in geophysics at the University of California, Berkeley, may represent a neutral voice. The evidence, he said, "is that CO_2 is just one of many influences."

(For Wikipedia, Mr. Rohde recently drew up graphic overviews of Phanerozoic carbon dioxide, en.wikipedia.org/wiki/Image:Phanerozoic-Carbon-Dioxide.png, and climate swings, en.wikipedia.org/wiki/Image:Phanerozoic-Climate-Change.png.)

For nearly two centuries, scientists have known that the ancient Earth went through ice ages and other climate upheavals. Their explanations included changes in land forms, ocean flows, solar intensity and Earth's orbit around the Sun.

The new argument dates from 1958, when scientists began to track carbon dioxide in the air, finding its levels low, 0.0315 percent, but increasing. They knew that excess gas could in theory trap more heat from the Sun, warming the planet and providing a new explanation for climate change.

The greenhouse theory rose to prominence in the 1980s as carbon dioxide continued to increase and as global temperatures started to increase. While scientists tracked many greenhouse gases, including ozone, methane and water vapor, they focused on carbon dioxide because its concentrations seemed to be rising quite rapidly.

Keen to put the threat in perspective, they sought to compare modern CO_2 levels to those of the past. Ice cores from the frozen regions turned out to harbor tiny air bubbles that showed carbon dioxide concentrations going back hundreds of thousands of years. Scientists found the preindustrial levels averaging 280 parts per million, down from 315 parts per million, or 0.0315 percent, in 1958.

Scientists suspected that the concentrations were once much higher, especially in hot eras of little or no polar ice. Eager to push beyond the cores, which went back just a half million years or so, scientists looked for ways to peer further back.

Dr. Berner of Yale focused on computer models. His studies of the Phanerozoic analyzed factors such as how some ages produced many volcanoes and much atmospheric carbon dioxide and others spawned mountains, extensive weathering of fresh rock and, by that mechanism, considerable uptake of atmospheric carbon dioxide.

From the start, he consistently reported close ties between carbon dioxide and climate swings. For instance, in the explosion of plant life from 400 million to 300 million years ago, he found a sharp drop in the gas, occurring as the earth entered an ice age.

"These results," Dr. Berner wrote in the journal *Science* in 1990, "support the notion that the atmospheric CO_2 greenhouse mechanism is a major control on climate over very long time scales."

Other scientists looked for clues among fossilized soils, plants and sea creatures, assuming that fluctuating climates had altered their growth patterns. In time, the ancient specimens yielded a bonanza of subtle evidence, some confirming aspects of Dr. Berner's modeling.

Claudia I. Mora and two colleagues at the University of Tennessee found that ancient soils verified the steep decline in carbon dioxide between 400 million and 300 million years ago.

Other scientists found conflicting evidence. In 1992, a team from the University of New Mexico reported that ancient soils showed extremely high levels of carbon dioxide 440 million years ago, an age of primitive sea life before the advent of land plants and animals. The carbon dioxide levels were roughly 16 times higher than today. Surprisingly, the scientists said, this appeared to coincide with wide glaciation, an analysis, wrote Crayton J. Yapp and Harald Poths in the journal *Nature*, that "suggests that the climate models require modification."

Throughout the 1990s, reconstruction papers offered evidence on both sides of the debate about the effects of carbon dioxide. Starting in 2000, the attacks intensified as Dr. Veizer of Ottawa questioned the CO_2-climate link across the whole Phanerozoic. He and two Belgian colleagues, writing in *Nature*, based their doubts on how two ice ages—440 million and 150 million years ago, in the age of dinosaurs—apparently had very high carbon dioxide levels.

In 2002, Daniel H. Rothman of the Massachusetts Institute of Technology also raised sharp Phanerozoic questions after studying carbon dioxide clues teased from marine rocks. Writing in *The Proceedings of the National Academy of Sciences*, he said that with one exception—the recent cool period of the last 50 million years—he could find "no systematic correspondence" between carbon dioxide and climate shifts.

In 2003, Dr. Veizer joined Nir J. Shaviv, an astrophysicist at the Hebrew University of Jerusalem, to propose a new climate driver. They envisioned slow movements of the solar system through the surrounding galaxy as controlling the cosmic rays that bombard Earth's atmosphere. A reduction, they argued, would lessen cloud cover and Earth's reflectivity, warming the planet. The reverse would cause cooling. The Phanerozoic record of cosmic-ray bombardment showed excellent agreement with climate fluctuations, trumping carbon dioxide, they wrote.

In 2004, Dr. Berner of Yale and four colleagues fired back. While saying cosmic rays were possibly "of some climatic significance," they argued that such an effect was much less than that of carbon dioxide.

In the debate, opponents can differ not only on the contours of past CO_2 fluctuations but also on defining hot and cold eras. Although Dr. Veizer sees a cold period 150 million years ago, a time of increased ice at sea but not on land because the continents had shifted from the poles, Dr. Berner, in his modeling, disregards it. Such differences can muddy the dispute.

Today, each side claims new victories. Dr. Veizer says he has a comprehensive paper on the cosmic-ray theory coming out soon. Dr. Berner recently refined his model to repair an old inconsistency.

The revision, described in the May issue of *The American Journal of Science*, brings the model into closer agreement with the fact of wide glaciation 440 million years ago, yielding what he sees as stronger evidence of the dominant role of carbon dioxide then.

Dr. Yapp, once a carbon dioxide skeptic, concurred, saying, "The data complied in the last decade suggests that long-term climate change correlates pretty well with CO_2 changes."

Some climatologists view the Phanerozoic debate as irrelevant. They say the evidence of a tie between carbon dioxide and planetary warming over the last few centuries is so compelling that any long-term evidence to the contrary must somehow be tainted. They also say greenhouse gases are increasing faster than at any other time in Earth history, making the past immaterial.

Carbon dioxide skeptics and others see the reconstructions of the last 15 years as increasingly reliable, posing fundamental questions about the claimed powers of carbon dioxide. Climatologists and policy makers, they say, need to ponder such complexities rather than trying to ignore or dismiss the unexpected findings.

"Some of the work has been quite meticulous," Thure E. Cerling, an expert at the University of Utah on Phanerozoic climates, said. "We are likely to learn something."

JONATHON KEATS

John Koza Has Built an Invention Machine

FROM *POPULAR SCIENCE*

Can creativity be automated? John Koza, whose "invention machine" was the recipient of the first U. S. patent awarded to a non-human, says yes. Jonathon Keats visits Koza's lab to investigate.

A S A HIGH-SCHOOL STUDENT in the 1950s, John Koza yearned for a personal computer. That was a tall order back then, as mass-produced data processors such as the IBM 704 were mainframes several times the size of his bedroom. So the cocksure young man went rummaging for broken jukeboxes and pinball machines, repurposing relays and switches and lightbulbs to make a computer of his own design.

Within certain parameters, his computer was a success, flawlessly reckoning the day of the week whenever he dialed in a calendar date, but the hardwiring made it useless for anything else. Koza's first invention was not about to supplant IBM, but the mothballed gizmo remains in his basement to this day, a reminder to himself that the intelligence of a machine is a matter of adaptability as much as accuracy.

Over the past several decades, Koza has internalized that lesson as deeply as any computer scientist alive and, arguably, made more of the insight than any coder in history. Now 62 and an adjunct professor at Stanford University, Koza is the inventor of genetic programming, a revolutionary approach to artificial intelligence (AI) capable of solving complex engineering problems with virtually no human guidance. Koza's 1,000 networked computers don't just follow a preordained routine. They create, growing new and unexpected designs out of the most basic code. They are computers that innovate, that find solutions not only equal to but better than the best work of expert humans. His "invention machine," as he likes to call it, has even earned a U.S. patent for developing a system to make factories more efficient, one of the first intellectual-property protections ever granted to a nonhuman designer.

Yet as impressive as these creations may be, none are half as significant as the machine's method: Darwinian evolution, the process of natural selection. Over and over, bits of computer code are, essentially, procreating. And over the course of hundreds or thousands of generations, that code evolves into offspring so well-adapted for its designated job that it is demonstrably superior to anything we can imagine. The age of creative machines has arrived. And its prophet is John Koza.

COMPUTER SCIENCE WAS STILL a brand-new discipline in the early 1960s, when Koza went to the University of Michigan. He was the second person anywhere to earn a bachelor's degree in the field. "I was interested in computers, so I studied computer science," he explains with characteristic bluntness. "Why do other people go into medicine or become policemen?" He earned his Ph.D. in December 1972, six months away from an academic job opportunity.

Industry, on the other hand, was eager for computer-science expertise. While in school, Koza had worked part-time for a supermarket rub-off game manufacturer called J&H International, calculating probabilities to keep game layouts unpredictable. A full-time position there was as good as certain. But the week of his graduation, the company shut its doors permanently.

Like many successful innovators, Koza combines unusual competence in his work with supreme confidence in himself. Rather than taking the bankruptcy as a sign that rub-off games were dead, he decided that scratch cards were the future of yet another moribund business: state lotteries. At the time, lotteries were weekly raffles, typically with six-digit tickets. A state might sell $1 million worth of tickets a week. Koza believed that, with a more interesting game, especially one offering instant gratification, he could sell more. He opened his own business with another former J&H employee and took a one-year gamble. By the end of 1974, they landed a contract with the state of Massachusetts for 25 million rub-off games.

The success of the instant-win lottery was, in a word, instantaneous—$2.7 million worth of tickets sold in the first week. "Our basic business was tripling lottery sales," Koza says. By 1982, dozens of states that didn't have a lottery had adopted one, and his company, Scientific Games, was diligently supplying most of them. Koza had invented a machine for printing money. He sold the company to Bally Manufacturing and ran the business under contract until 1987. At which point he found himself very rich but, for the second time in his life, jobless.

JOHN KOZA HOVERS over a computer terminal in a cramped office a mile from the building where he keeps his invention machine. Seated at the terminal is a clean-cut researcher named Lee Jones, one of Koza's two employees. His other employee, Sameer Al-Sakran, leans over a second terminal, stroking his facial stubble.

Jones is reviewing one of the invention machine's latest accomplishments, which Koza is preparing to present at the annual Genetic and Evolutionary Computation Conference, familiarly called GECCO. In this instance, the machine has created a complex lens system that outperforms a wide-field eyepiece for telescopes and binoculars patented just six years ago by lens designers Noboru Koizumi and Naomi Watanabe—and which does so, moreover, without infringing on the Koizumi-Watanabe patent.

Jones calls up an optical simulator known as KOJAC. From a prescription (which numerically describes the curvature, thickness and glass type of lens components), KOJAC predicts how the compound lens will function in the real world. The numerous variables make the effect of simple changes difficult to predict. As a result, lens designers are a creative bunch, who depend as heavily on intuition as on knowledge.

What Koza has done is to automate the creative process. To begin, the invention machine randomly generates 75,000 prescriptions. It then analyzes them in KOJAC, which assigns each a fitness rating based on how close it comes to a desired set of specifications—in this case, a wide field of view with minimal distortion. None of the 75,000 members of the first generation will be usable wide-field telescopic eyepieces. But a few of these primitive systems will be marginally effective at focusing a wide field of view, and a couple others might slightly reduce distortion in one way or another.

From there, it's Darwinism 101. The invention machine mates some systems together, redistributing characteristics from two parent lens systems into their offspring. Others it mutates, randomly altering a single detail. Other lenses pass on to the next generation unchanged. And then there are the cruel necessities of natural selection: The machine expels most lenses with low fitness ratings from the population, kills them off so their genetic material won't contaminate the others.

Koza asks Jones to pull up the stats on the wide-field telescopic eyepiece. Amid a rush of figures, he reads off the number "295." That's how many generations it took for genetic programming to engineer around the Koizumi-Watanabe patent. In fact, the invention machine's lens is *better* than the Koizumi-Watanabe system: Because it keeps breeding until all design specs are met, often some performance requirements are exceeded by the end of the run. The final field-of-view for Koza's eyepiece is a remarkable 10 degrees higher than the 55 degrees achieved by Koizumi and Watanabe.

Jones swiftly rotates through several other recent inventions, all generated using the same technique as the lens system. There are logic circuits and amplifiers and filters, some of them suitable for the challenging low-power requirements of cellphones and laptops. Each took between one day and one month to evolve, generating an electricity bill of more than $3,000 a month.

LIKE EVERY ENGINEERING BREAKTHROUGH, genetic programming did not emerge fully formed from the ether. Rather it grew out of two promising yet unfulfilled lines of research in computer science: genetic algorithms and artificial intelligence.

Koza's thesis adviser at the University of Michigan was John Holland, the man widely regarded as the father of genetic algorithms. While Holland was a grad student studying mathematics at Michigan in the 1950s, he'd happened upon a book called *The Genetical Theory of Natural Selection*, written by English biologist Ronald Fisher in 1930. The book laid out, in strict mathematical terms, the basic mechanism of variation in plants and animals. "I thought I'd try to figure out a program that did that," Holland recalls. He envisioned a system that, through small, incremental improvements, would breed good code the way a farmer breeds good corn—an early forerunner to genetic programming in the sense that, say, Mendelian inheritance was the first step toward understanding Darwinian evolution.

Holland and his student David Goldberg implemented the idea

in 1980. Goldberg had studied civil engineering before working on his doctorate and was interested in the practical problem of how computers could be used to optimize the capacity of gas pipelines. They began by creating a rough model of an efficient layout for the pipeline. The software then made small random changes to the system—alternately varying the pressure, flow rate and pumping schedule—and simulated the gas flow anew following each mutation cycle. The computer retained any alteration that improved performance, even by the slightest bit, and discarded all those that didn't. Over 20 or 30 generations, the system evolved, by almost imperceptible steps, to become markedly better than it was at the start.

The power of genetic algorithms increased in step with the power of the processors they ran on. By the mid–1980s, Holland's process had spawned a small cottage industry, complete with dedicated academic conferences and myriad industrial applications.

Artificial intelligence, on the other hand, was decidedly less practical then. Its goal was (and largely remains) to model human cognitive functions, such as language use and pattern recognition, in computer systems—to make machines think. AI researchers were also hugely optimistic. (One conference topic, seriously debated, was "Should AI run for president?")

Koza had just left the lottery business. He was interested in the commercial potential of genetic algorithms and thought, given his financial success and the strong economy, that he would become a venture capitalist. He read Holland's book on genetic algorithms, subscribed to journals, and attended meetings, all of which reminded him how much he had enjoyed pure research as a grad student and how much he missed it. "I became more and more interested in the technical problems," he recalls. "I realized that venture capital was just another hectic business."

He took a position at Stanford as an adjunct professor, which gave him time to absorb the latest research in genetic algorithms and artificial intelligence. Yet he felt that both disciplines were somehow fundamentally lacking. Pragmatic by nature, Koza was frustrated—as many others would later become—by the growing gap between promise and performance in artificial intelligence. Meanwhile genetic algorithms presented a different kind of frustration: Although they proved to be excellent optimizers, perfect for tweaking well-defined systems, they lacked the creative capability to come up with novel solutions to their problems.

In 1987 Koza was on an airplane, returning to California from an AI conference in Italy, when he had the crucial insight that Holland himself would later deem revolutionary. AI was all promise, an underachieving prodigy. Genetic algorithms were all performance, reliable drones. Koza was 30,000 feet above Greenland when he asked himself why a genetic algorithm, so adept at refining pipelines, couldn't be used to evolve its own software. Why couldn't a computer program adapt *itself* and, in doing so, solve any problem fed into it?

The key was mating bits of computer programs together, not just strings of numbers. The old genetic algorithms worked to optimize specific parameters; Koza's leap in genetic programming allowed for open-ended evolutions of basic structure and so produced more novel and sophisticated designs. "If you're trying to breed a better racehorse, you could take a herd of horses out into a field and irradiate them and hope that, through random mutation, you get a better horse," Koza explains. "Or you could mate a champion with another champion."

His breakthrough moment came in October 1995. Working with only the most rudimentary information, he watched his computer evolve a circuit. Unlike the gas pipelines bred by genetic algorithms, his circuit didn't start with an inferior design that was optimized. Beginning from a pile of unconnected components—assorted resistors, capacitors and the like—his computer devised a complex electronic circuit.

Koza made a drawing of his new circuit and showed it to a colleague, who told him that it was a low-pass filter—a circuit used for cleaning up the signal passing through an amplifier. He also learned that someone had taken out a patent on it, which made him wonder whether genetic programming could evolve other patented circuits from scratch. He programmed the machine to design circuits with the attributes of other patented devices and started churning out infringements by the dozen. "That's when we began to see that genetic programming could be human-competitive," he says. "If you remake a patented circuit, you're doing something that people consider inventive."

JANUARY 25, 2005, looms large in the history of computer science as the day that genetic programming passed its first real Turing test: The examiner had no idea that he was looking at the intellectual property of a computer. This is especially significant because the U.S. Patent and Trademark Office requires a "non-obvious step"—a break from established practices or what someone might deduce from them—to grant an invention intellectual-property protection. The machine was demonstrably creative.

And that was just the start. Every day now, genetic programs continue to create the unexpected, the counterintuitive or the just plain weird. Take, for example, the antenna that was set to launch on NASA's Space Technology 5 mission, a test platform for new technologies. Several years ago, Koza protégé Jason Lohn took a job as a computer scientist at NASA Ames Research Center, which had taken an interest in evolutionary approaches to problem solving, including genetic algorithms. "Antenna design has a black-art quality to it," Lohn says. "A lot of it is intuition."

Lohn got his hands on the antenna specs for the Space Technology 5 mission. He plugged in an antenna's basic requirements and let the software run. What he got, several hundred generations later, appeared to be a mistake. "It looked like a bent paper clip," he remembers. Lohn had no background in antenna design, but that was beside the point. "There's no chapter in the textbooks on crooked wire antennas," he says. He had his bent paper clip prototyped and put it in a test chamber. Sure enough, it provided the tricky combination of wide bandwidth and wide beam that NASA required. Like the duck-billed platypus, it looked preposterous but proved perfectly suited to its niche.

John Holland has lately been researching what such ingenuity might tell us about the creative process in humans. He believes that revolutionary ideas don't come at random but are "new combinations of fairly standard parts with which we're already familiar." He cites as examples the internal combustion engine and the airplane, for which all the components were available long before the invention came along, lacking only someone with adequately broad knowledge, deep resources and the temperament to combine them. "Evolution is good at recombining building blocks to get innovations," Holland says. The machine has inspired a new way to think about our own creative process: Perhaps extraordinary thinking is simply the product of gradual refinements and serendipitous recombinations. Darwin's combination of mutation, sex and selection creates not just new species, or antennas: It spawns creativity itself.

AS GENETIC PROGRAMMING becomes pervasive over the next decade, the process of finding good solutions to difficult engineering problems will become efficient in the way that once-arduous tasks such as 3-D rendering have become routine. But, as with 3-D rendering, the real challenge will lie in deciding what to create. This is no trivial matter—to invent a car, even with an invention machine, you must be able to conceive of wanting a horseless carriage in the first place. In the future, as solutions become plentiful and cheap, the real test of creativity will come in the search for problems.

Meanwhile, Koza thinks he has already found a good one to solve, although it will require more than just an invention machine. Confident as ever in the power of rational thought (and in himself), he has undertaken the mission of reengineering U.S. elections. More specifically, he has developed a strategy to effectively eliminate the American electoral college. Here's how it works: Since the Constitution allows each state to assign its electors any way it chooses, a critical mass of states could pledge to assign its electors to vote for the winner of the nationwide popular vote. In theory, as few as 11 states could make this work. The plan is slowly gaining traction among politicians and pundits, and a bill was recently introduced in the Illinois legislature that would make that state the first to back the plan.

As Koza waits for the machinations of politics to grind, he has his sights set on one more goal, perhaps genetic programming's ultimate Turing test: an invention that succeeds in the marketplace. (Passing muster with a patent examiner is nothing compared to nabbing a couple million customers.) That means finding a problem serious enough that people will pay to have it solved. Alas, figuring out what people really need is one thing genetic programming can't do. Koza will have to invent that for himself.

JOHN CASSIDY

Mind Games

FROM *THE NEW YORKER*

Economists have long assumed that people make rational choices in financial decisions. Now, with brain-imaging, a new breed of neuro-economists can put that assumption to the test. John Cassidy finds out what it's like to say, "This is your brain on money."

LIKE MANY PEOPLE who have accumulated some savings, I invest in the stock market. Most of my retirement money is invested in mutual funds, but now and again I also buy individual stocks. My holdings include the oil company Royal Dutch Shell, the drug company GlaxoSmithKline, and the phone company British Telecommunication. I like to think that I picked these stocks because I can discern value where others can't, but my record hardly backs this up. I invested in BT in 2001, shortly after the Nasdaq crashed, when the stock had already fallen substantially, only to watch it slide another fifty per cent. I should have sold out, but I held on, hoping for a rebound. Five years later, the stock is trading well below the price I paid for it, and I still own it.

I sometimes wonder what goes on in my head when I make stupid investment decisions. A few weeks ago, I had a chance to find out, when I took part in an experiment at New York University's Center for Brain Imaging, in a building off Washington Square Park. In the lobby, I met Peter Sokol-Hessner, a twenty-four-year-old graduate student, who escorted me to a control room full of computers. Sokol-Hessner is completing a doctorate in psychology, but he is currently working on a research project in the emerging field of neuroeconomics, which uses state-of-the-art imaging technology to explore the neural bases of economic decision-making.

Sokol-Hessner is particularly interested in "loss aversion," which is what I was suffering from when I refused to sell my BT stock. During the past decade or so, economists have devised a series of experiments to demonstrate just how much we dislike losing money. If you present people with an even chance of winning a hundred and fifty dollars or losing a hundred dollars, most refuse the gamble, even though it is to their advantage to accept it: if you multiply the odds of winning—fifty per cent—times a hundred and fifty dollars, minus the odds of losing—also fifty per cent—times a hundred dollars, you end up with a gain of twenty-five dollars. If you accepted this bet ten times in a row, you could expect to gain two hundred and fifty dollars. But, when people are presented with it once, a prospective return of a hundred and fifty dollars isn't enough to compensate them for a possible loss of a hundred dollars. In fact, most people won't accept the gamble unless the winning stake is raised to two hundred dollars.

Why are we so averse to losses, even at the expense of gains? At the Center for Brain Imaging, I removed my belt and shoes and entered a room containing a big metal box, which measured about six feet by six feet by six feet, with a slim gurney protruding from one side. It was a magnetic-resonance-imaging machine, identical to those hospitals use to scan bodies for lesions and tumors. "As blood pumps through the brain, the oxygen it contains causes small changes in the magnetic field," Sokol-Hessner explained. "The scanner can pick up on that and tell us where the blood is flowing. We get a picture of which parts of the brain are being used."

I put on earplugs and lay back on the gurney. Sokol-Hessner and two lab assistants placed some foam around my ears and lowered a plastic grille over my face. In one of my hands they placed a metal console with two buttons on it. I felt my head and shoulders sliding into a long, cylindrical hole about a foot and a half wide. "Take a few deep breaths," Sokol-Hessner said. There was a crashing noise—the sound of the magnet warming up. Struggling to fend off claustrophobia, I closed my eyes and counted to a hundred while the scanner took a picture. "How are you doing?" asked Sokol-Hessner, who had retreated to the control room. "Fine," I lied.

My task was to consider a series of investment options that were presented on a small illuminated screen over my head. In each case, one of the options would be a fifty-fifty bet and the other would be a sure thing. The first scenario appeared on the screen: a possible gain of four dollars and a possible loss of two dollars versus a sure thing of zero, meaning that I wouldn't win or lose anything. I had three seconds to make my selection. Two dollars didn't seem like a lot to lose, so I pressed a button on the console to accept the bet. Somewhere in the next room, a random number generator was deciding whether I had won or lost. Then this message flashed on the screen: "You won $4.00."

SOKOL-HESSNER'S THESIS ADVISERS are Elizabeth Phelps, a professor of psychology and neural science at N.Y.U., and Colin Camerer, an economist at Caltech who helped found neuroeconomics. This spring, I visited Camerer at his office in Pasadena, California. He is a stocky man of forty-six, with a large, bald head and blue eyes. His office was cluttered with textbooks and academic journals, and on one wall there was a whiteboard covered with equations. It looked like every other economist's office I've visited, except that on Camerer's desk there was a plastic model of the human brain.

While we were speaking, Camerer picked up the model and gave me a quick tour, starting at the front, with the prefrontal cortex, a structure that helps us perform complicated mental tasks, such as logical reasoning and planning. Then he pointed to the parietal cortex and the temporal lobes, regions that are also involved in deliberative decision-making. All these areas are much larger in humans than in other animals; scientists think that they were the last parts of the brain to evolve.

The model was made of layers of interlocking pieces. Camerer removed a piece from the top layer, exposing the so-called limbic areas beneath, including the insular cortex and the striatum. These structures date to the earliest period of human evolution, and neuroscientists believe that they help us process emotions. Camerer was particularly eager to show me the amygdala, a pair of almond-shaped structures that also play a role in the processing of emotions. "They are in here somewhere," he said, removing more pieces from the model.

Camerer was a child prodigy. He grew up in Baltimore and entered college at Johns Hopkins at the age of fourteen, majoring in mathematics. He spent a lot of time at a local racetrack, betting on horses, a hobby that got him interested in risk-taking and decision-making. In 1981, when he was twenty-one, he obtained a Ph.D. in economics at the University of Chicago Graduate School of Business. Camerer also found inspiration outside his field.

In 1979, two Israeli psychologists, Daniel Kahneman and Amos Tversky, published a paper in the economics journal *Econometrica*, describing the concept of loss aversion. At the time, few economists and psychologists talked to one another. In the nineteenth century, their fields had been considered closely related branches of the "moral sciences." But psychology evolved into an empirical discipline, grounded in close observation of human behavior, while economics became increasingly theoretical—in some ways it resembled a branch of mathematics. Many economists regarded psychology with suspicion, but their preference for abstract models of human behavior came at a cost.

In order to depict economic decisions mathematically, economists needed to assume that human behavior is both rational and predictable. They imagined a representative human, *Homo economicus*, endowed with consistent preferences, stable moods, and an enviable ability to make only rational decisions. This sleight of hand yielded some theories that had genuine predictive value, but economists were obliged to exclude from their analyses many phenomena that didn't fit the rational-actor framework, such as stock-market bubbles, drug addiction, and compulsive shopping. Economists continue to study *Homo economicus*, but many recognize his limitations. Over the past twenty-five years, using methods and insights borrowed from psychology, they have devised a new approach to studying decision-making: behavioral economics.

One of Camerer's mentors, Richard Thaler, was among the first economists to cite Kahneman and Tversky's work; beginning in 1987, he published a series of influential articles describing various types of apparently irrational behavior, including loss aversion.

Acknowledging that people don't always behave rationally was an important, if obvious, first step. Explaining *why* they don't has proved much harder, and recently Camerer and other behavioral economists have turned to neuroscience for help. By the mid-nineteen-nineties, neuroscientists, using MRI machines and other advanced imaging techniques, had developed a basic understanding of the roles played by different parts of the brain in the performance of particular tasks, such as recognizing visual patterns, doing mental computations, and reacting to threats. In the mid-nineties, Antonio Damasio, a neurologist at the University of Iowa, and Joseph LeDoux, a neuroscientist at N.Y.U., each published a book for lay readers describing how the brain processes emotions. "We were reading the neuroscience, and it just seemed obvious that there were applications to economics, both in terms of ideas and methods," said George Loewenstein, an economist and psychologist at Carnegie Mellon who read Damasio's and LeDoux's books. "The idea that you can look inside the brain and see what is happening is just so intensely exciting."

In 1997, Loewenstein and Camerer hosted a two-day conference in Pittsburgh, at which a group of neuroscientists and psychologists gave presentations to about twenty economists, some of whom were inspired to do imaging studies of their own. In the past few years, dozens of papers on neuroeconomics have been published, and the field has attracted some of the most talented young economists, including David Laibson, a forty-year-old Harvard professor who is an expert in consumer behavior. "Natural science has moved ahead by studying progressively smaller units," Laibson told me. "Physicists started out studying the stars, then they looked at objects, molecules, atoms, subatomic particles, and so on. My sense is that economics is going to follow the same path. Forty years ago, it was mainly about large-scale phenomena, like inflation and unemployment. More recently, there has been a lot of focus on individual decision-making. I think the time has now come to go beyond the individual and look at the inputs to individual decision-making. That is what we do in neuroeconomics."

WHEN PEOPLE MAKE INVESTMENTS, they weigh the possible outcomes of their decisions and select a portfolio of stocks and bonds that offers the highest possible return at an acceptable level of risk. That is what mainstream economics says, anyway. In fact, people often have only a vague idea of the risks they face. Consider my investment in BT. Back in 2002, there was no way that I could have predicted how much profit the company would make in 2006, let alone in 2010 or 2020. I bought the stock, nonetheless, convinced that it could only increase in value.

As imaging technology gets more sophisticated and easier to use, it may become possible to monitor investors' brains while they trade stocks at their offices. For now, however, economists are restricted to laboratory experiments, in which they pay volunteers to play simple games designed to imitate situations that people experience in daily life. In one study, Camerer and several colleagues performed brain scans on a group of volunteers while they placed bets on whether the next card drawn from a deck would be red or black. In an initial set of trials, the players were told how many red cards and black cards were in the deck, so that they could calculate the probability of the next card's being a certain color. Then a second set of trials was held, in which the participants were told only the total number of cards in the deck.

The first scenario corresponds to the theoretical ideal: investors facing a set of known risks. The second setup was more like the real world: the players knew something about what might happen, but not very much. As the researchers expected, the players' brains reacted to the two scenarios differently. With less information to go on, the players exhibited substantially more activity in the amygdala and in the orbitofrontal cortex, which is believed to modulate activity in the amygdala. "The brain doesn't like ambiguous situations," Camerer said to me. "When it can't figure out what is happening, the amygdala transmits fear to the orbitofrontal cortex."

The results of the experiment suggested that when people are confronted with ambiguity their emotions can overpower their reasoning, leading them to reject risky propositions. This raises the intriguing possibility that people who are less fearful than others might make better investors, which is precisely what George Loewenstein and four other researchers found when they carried out a series of experiments with a group of patients who had suffered brain damage.

Each of the patients had a lesion in one of three regions of the brain that are central to the processing of emotions: the amygdala, the orbitofrontal cortex, or the right insular cortex. The researchers presented the patients with a series of fifty-fifty gambles, in which they stood to win a dollar-fifty or lose a dollar. This is the type of gamble that people often reject, owing to loss aversion, but the patients with lesions accepted the bets more than eighty per cent of the time, and they ended up making significantly more money than a control group made up of people who had no brain damage. "Clearly, having frontal damage undermines the over-all quality of decision-making," Loewenstein, Camerer, and Drazen Prelec, a psychologist at M.I.T.'s Sloan School of Management, wrote in the March, 2005, issue of the *Journal of Economic Literature*. "But there are situations in which frontal damage can result in superior decisions."

NOT LONG AGO, I drove to Princeton University to speak to Jonathan Cohen, a fifty-year-old neuroscientist who is the director of Princeton's Center for the Study of Brain, Mind, and Behavior. Nine years earlier, while he was teaching at Carnegie Mellon, Cohen attended the conference that Camerer and Loewenstein organized. "I had never taken any economics courses; I had no idea what they did," he recalled. "I thought it was all about setting interest rates."

Since then, Cohen has collaborated with economists on several imaging studies. "The key idea in neuroeconomics is that there are multiple systems within the brain," Cohen said. "Most of the time, these systems coöperate in decision-making, but under some circumstances they compete with one another."

A good way to illustrate Cohen's point is to imagine that you and a stranger are sitting on a park bench, when an economist approaches and offers both of you ten dollars. He asks the stranger to suggest how the ten dollars should be divided, and he gives you the right to approve or reject the division. If you accept the stranger's proposal, the money will be divided between you accordingly; if you refuse it, neither of you gets anything.

How would you react to this situation, which economists refer to as an "ultimatum game," because one player effectively gives the other an ultimatum? Game theorists say that you should accept any positive offer you receive, even one as low as a dollar, or you will end up with nothing. But most people reject offers of less than three dollars, and some turn down anything less than five dollars.

Cohen and several colleagues organized a series of ultimatum games in which half the players—the respondents—were put in MRI machines. At the beginning of a round, each respondent was shown a photograph of another player, who would make the respondent an offer. The offer then appeared on a screen inside the MRI machine, and the respondent had twelve seconds in which to accept or reject it. The results were the same as in other, similar experiments—low offers were usually vetoed—but the respondents' brain scans were revealing.

When respondents received stingy offers—two dollars for them, say, and eight dollars for the other player—they exhibited substantially more activity in the dorsolateral prefrontal cortex, an area associated with reasoning, and in the bilateral anterior insula, part of the limbic region that is active when people are angry or in distress. The more activity there was in the limbic structure, the more likely the person was to reject the offer. To the researchers, it looked as though the two regions of the brain might be competing to decide what to do, with the prefrontal cortex wanting to accept the offer and the insula wanting to reject it. "These findings suggest that when participants reject an unfair offer, it is not the result of a deliberative thought process," Cohen wrote in a recent article. "Rather, it appears to be the product of a strong (seemingly negative) emotional response."

Several explanations have been proposed for people's visceral reaction to unfair offers. Maybe human beings have an intrinsic preference for fairness, and we get angry when that preference is violated—so angry that we punish the other player even at a cost to ourselves. Or perhaps people reject low offers because they don't want to appear weak. "We evolved in small communities, where there was a lot of repeated interaction with the same people," Cohen said. "In such an environment, it makes sense to build up a reputation for toughness, because people will treat you better next time they see you."

Unfortunately, some of the emotional responses that we developed millennia ago no longer serve us well. As Cohen put it, "Does it make sense to play tough with a person you meet on a street in L.A.? No. For one thing, you will probably never see that person again. For another, he may pull out a gun and shoot you." Obviously, we can't alter our brain structures, but it may be possible to influence decision-making by tinkering with brain chemistry. Last year, a group of economists led by Ernst Fehr, of the University of Zurich, demonstrated how this might be done, in an experiment involving what economists call "the trust game."

Trust plays a key role in many economic transactions, from buying a secondhand car to choosing a college. In the simplest version of the trust game, one player gives some money to another player, who invests it on his behalf and then decides how much to return to him and how much to keep. The more the first player invests, the more he stands to gain, but the more he has to trust the second player. If the players trust each other, both will do well. If they don't, neither will end up with much money.

Fehr and his collaborators divided a group of student volunteers into two groups. The members of one group were each given six puffs of the nasal spray Syntocinon, which contains oxytocin, a hormone that the brain produces during breast-feeding, sexual intercourse, and other intimate types of social bonding. The members of the other group were given a placebo spray.

Scientists believe that oxytocin is connected to stress reduction, enhanced sociability, and, possibly, falling in love. The researchers hypothesized that oxytocin would make people more trusting, and their results appear to support this claim. Of the twenty-nine students who were given oxytocin, thirteen invested the maximum money allowed, compared with just six out of twenty-nine in the control group. "That's a pretty remarkable finding," Camerer told me. "If you asked most economists how they would produce more trust in a game, they would say change the payoffs or get the participants to play the game repeatedly: those are the standard tools. If you said, 'Try spraying oxytocin in the nostrils,' they would say, 'I don't know what you're talking about.' You're tricking the brain, and it seems to work."

ECONOMICS HAS ALWAYS BEEN CONCERNED with social policy. Adam Smith published *The Wealth of Nations*, in 1776, to counter what he viewed as the dangerous spread of mercantilism; John Maynard Keynes wrote *The General Theory of Employment, Interest, and Money* (1936) in part to provide intellectual support for increased government spending during recessions; Milton Friedman's *Capitalism and Freedom*, which appeared in 1962, was a free-market manifesto. Today, most economists agree that, left alone, people will act in their own best interest, and that the market will coördinate their actions to produce outcomes beneficial to all.

Neuroeconomics potentially challenges both parts of this argument. If emotional responses often trump reason, there can be no presumption that people act in their own best interest. And if markets reflect the decisions that people make when their limbic structures are particularly active, there is little reason to suppose that market outcomes can't be improved upon.

Consider saving for retirement. Surveys show that up to half of all families end their working lives with almost no financial assets, other than their entitlement to Social Security benefits. Saving money is difficult, because it involves giving up things that we value now—a new car, a vacation, fancy dinners—in order to secure our welfare in the future. All too often, the desire for immediate gratification prevails. "We humans are very committed to our long-term goals, such as eating healthy food and saving for retirement, and yet, in the moment, temptations arise that often trip up our long-term plans," David Laibson, the Harvard economist, said. "I was planning to give up smoking, but I couldn't resist another cigarette. I was planning to be faithful to my wife, but I found myself in an adulterous relationship. I was planning to save for retirement, but I spent all my earnings. Understanding this tendency stands at the heart of a lot of big policy debates."

Laibson has collaborated with Loewenstein, Cohen, and Samuel

McClure, another Princeton psychologist, to examine what happens in people's brains when they are forced to choose between immediate and delayed rewards. For a study the four researchers published in *Science*, in 2004, they used an MRI machine to scan a group of student volunteers who were asked to choose between receiving a fifteen-dollar Amazon.com gift voucher today and receiving a twenty-dollar Amazon.com gift voucher in two weeks or a month.

The scans showed that both gift options triggered activity in the lateral prefrontal cortex, but that the immediate option also caused disproportionate activity in the limbic areas. Moreover, the greater the activity in the limbic areas the more likely the students were to choose the voucher that was immediately available and less valuable.

The results provide further evidence that reason and emotion often compete inside the brain, and it also helps explain a number of puzzling phenomena, such as the popularity of Christmas savings accounts, which people contribute to throughout the year. "Why would anybody put money into a savings account that offers zero interest and imposes a penalty if you withdraw cash early?" Cohen said. "It simply doesn't make sense in terms of a traditional, rational economic model. The reason is that there is this limbic system that produces a strong drive. When it sees something it likes, it wants it now. So you need some type of pre-commitment device to make people save."

Laibson and Brigitte Madrian, an economist at the Wharton School, have studied one such "pre-commitment device" for 401(k) plans, which deduct part of an employee's earnings each month and invest them in stocks and bonds. Because the plans are often optional, many people fail to join them, even when their employers offer to match a portion of their contributions. Laibson and his colleagues have called for people to be automatically included in the plans unless they choose to opt out. At companies that have adopted such a policy, enrollment rates have increased sharply.

Reforming 401(k) plans is an example of "asymmetric paternalism," a new political philosophy based on the idea of saving people from the vagaries of their limbic regions. Warning labels on tobacco and potentially harmful foods are similarly intended to keep subcortical structures in check. Neuroeconomists have suggested additional policies, including warning buyers of lottery tickets that their chances of winning are practically nonexistent and imposing mandatory "cooling off" periods before people make big-ticket purchases, such as cars and boats. "Asymmetric paternalism helps those whose rationality is bounded from making a costly mistake and harms more rational folks very little," Camerer, Loewenstein, and three colleagues wrote in a 2003 issue of the *University of Pennsylvania Law Review* "Such policies should appeal to everyone across the political spectrum."

SOME NEUROECONOMIC "FINDINGS" aren't exactly discoveries, of course. In the fourth century B.C., Plato described reason as a charioteer attempting to steer the twin horses of passion and spirit. More recently, Freud wrote about the contest between the ego and the id. "What is new," Jonathan Cohen wrote in the fall, 2005, issue of the *Journal of Economic Perspectives*, "is that researchers now have the tools to begin to identify and characterize these systems at the level of their physical implementation in the human brain. Neuroscience gives detailed access to the mechanisms that underlie behavior and thus may allow scientists to answer questions that cannot be answered easily, or at all, by observing behavior alone."

Many traditional economists are unimpressed by this argument. In a recent paper, "The Case for Mindless Economics," Faruk Gul and Wolfgang Pesendorfer, two Princeton economists, wrote, "Neuroscience evidence cannot refute economic models because the latter make no assumptions and draw no conclusions about the physiology of the brain." Gul and Pesendorfer have a point: neuroeconomics doesn't tell us whether the neo-Keynesian or the neoclassical model of inflation is correct. But it can provide indirect evidence to reinforce certain theories and discredit others. About ten years ago, David Laibson published a paper on "hyperbolic discounting," which suggested that people treat immediate rewards differently from the way they treat delayed rewards, preferring the former in a manner that simple rational-choice models can't explain. Now the results of the Amazon voucher experiment have provided a possible explanation for the behavior that Laibson identified: immediate and delayed rewards stimulate different parts of the brain. "The practical implications of the experiment come from obtaining a better understanding of the human taste for instant gratification," Laibson said. "If we can understand that, we will be in a much better position to design policies that mitigate what can be self-defeating behavior."

The biggest challenge facing neuroeconomics comes not from its opponents in the economics profession but from its supposed allies in neuroscience. Many neuroscientists now consider MRI data to be uninformative. Neural activity occurs in milliseconds, on a scale of perhaps 0.1 millimetres. A typical MRI machine, which measures neural firing indirectly, by tracking blood flow, takes a picture every couple of seconds and isn't able to detect anything less than three millimetres long. Because of these limitations, neuroscientists prefer to track the firing of single neurons by inserting tiny electrodes into the brain. Unfortunately, this is an invasive procedure, and its experimental use has generally been restricted to laboratory animals.

There is also a more fundamental objection to neuroeconomics and the Platonic view of decision-making. "There is no evidence that hidden inside the brain are two fully independent systems, one rational and one irrational," Paul W. Glimcher, a neuroscientist who is the director of N.Y.U.'s Center for Neuroeconomics, and two of his colleagues, Michael C. Dorris and Hannah M. Bayer, wrote in a recent paper. "There is, for example, no evidence that there is an emotional system, per se, and a rational system, per se, for decision making at the neurobiological level."

In place of the reason-versus-passion model, Glimcher and his colleagues have adopted a view of decision-making that, paradoxically, bears a striking resemblance to orthodox economics. In one experiment, Glimcher and a colleague trained thirsty monkeys to direct their eyes to one of two illuminated targets, which earned them differing chances of getting juice rewards—a fifty-per-cent chance of getting a full cup of juice for looking right, say, versus a seventy-per-cent chance of getting half a cup of juice for looking left. The game was repeated many times, with the probabilities changing periodically.

The monkeys' task was to consume as much juice as possible, and they proved very adept at it. Before long, they were dividing their time between the illuminated targets in a way that roughly maximized their payoffs. When the odds favored looking right, they looked right; when the odds favored looking left, they looked left. Glimcher also used electrodes to track neural firing in part of the posterior parietal cortex, an area that is thought to organize signals transmitted by the retina. He discovered that the firing rate was closely related to the rewards the monkeys were likely to receive. "Specifically," he and his colleagues reported, "the firing rate of a neuron associated with a leftward movement was a linear function of the probability that the leftward movement would yield the juice reward."

Clearly, monkeys can't do probability sums. (Many humans struggle with them!) But Glimcher's experiment implies that their brains act as if they were solving a mathematical problem, which is what economists assume when they depict people as rational agents trying to maximize their well-being, or "utility." "What seems to be emerging from these early studies is a remarkably economic view of the primate brain," Glimcher and his colleagues wrote. "The final stages of decision-making seem to reflect something very much like a utility calculation."

If Glimcher's results could be demonstrated in human brains, they might undermine a lot of neuroeconomics, and many in the field tend to downplay his work. "Well, monkeys are very interesting, but they are not nearly as rich in their behavior as humans," George Loewenstein said to me. "Humans have this very well-developed prefrontal cortex, which allows us to look ahead a number of stages, rather than just behaving in a reflexive fashion. Still, it's wonderful that we have these controversies. Most of us are friends, and we debate these issues. I've learned a lot from talking to Paul."

I WAS INSIDE THE MRI MACHINE for nearly two hours, and I answered more than two hundred and fifty questions, which were organized into two blocks. Sokol-Hessner had instructed me to answer the first set as if each investment were the only one I would make. He told me to treat the second set of gambles as a group, as if I were constructing an investment portfolio. Later, he explained that he wanted to compare my answers to the two blocks of questions. Many people become less loss-averse when they are constructing a portfolio of investments, presumably because they believe that losses in one part of their portfolio will be made up for by gains in others. "Our research has shown that people can alter their own choice behavior in a systematic fashion," Sokol-Hessner said. "They can make themselves less loss-averse. If loss aversion is mediated by the limbic structures, such as the amygdala, we would expect a big decrease in activity in those areas when you become less loss-averse."

The goal of the imaging experiment was to test this hypothesis. I was only the second person to take part in the experiment, but Sokol-Hessner told me that I was an atypical case. Rather than altering my strategy, I answered all the questions in the same way. Whenever the risk-free option was worth more than about five dollars, I accepted it, thinking that I would have been foolish to turn down a sure thing. Occasionally, when the risk-free option was zero, or close to zero, I gambled on the risky option. I'm not sure why I acted in this way—it wasn't strictly logical—but it made answering the questions easy, and it seemed to pay off: by the end of the experiment, I had won sixty-eight dollars.

My experience illustrated some of the drawbacks of brain scanning. After about an hour inside the machine, I was more concerned about getting out than I was about making a few dollars. (Sokol-Hessner said that I moved my head around so much that my brain scans were unusable.) "That's the terrible thing about MRIs," Sokol-Hessner conceded. "You are in a long tube, and you might well feel tired or claustrophobic. There's definitely other stuff going on in there besides the experiment. We have to be very careful about how we interpret the evidence."

Economists who have staked their careers on neuroeconomics are mindful of this advice. "It isn't a wholesale rejection of the traditional methodology," David Laibson said of his field. "It is just a recognition that decision-making is not always perfect. People try to do the best they can, but they sometimes make mistakes. The idea that a single mechanism maximizes welfare and always gets things right—that concept is on the rocks. But models that I call 'cousins' of the rational-actor model will survive."

The modified theories to which Laibson referred assume that people have two warring sides: the first deliberative and forward-looking, the second impulsive and myopic. Under certain circumstances, the impulsive side prevails, and people succumb to things like drug addiction, overeating, and taking wild gambles in the stock market. For now, the new models await empirical verification, but neuroeconomists are convinced that they're onto something. "We are not going to falsify all of traditional economics," Colin Camerer said. "But we are going to point to a whole range of biological variables that traditionally have not been included in the analysis. In economics, that is a big change."

BARRY YEOMAN

Schweitzer's Dangerous Discovery

FROM *DISCOVER*

> *Recently the field of paleontology was roiled by dispute when Mary*
> *Schweitzer discovered remnants of tissue in dinosaur bones—some-*
> *thing that had once been considered impossible. Barry Yeoman pro-*
> *files the unconventional scientist at the center of the controversy.*

EVER SINCE MARY HIGBY SCHWEITZER peeked inside the fractured thighbone of a *Tyrannosaurus rex,* the introverted scientist's life hasn't been the same. Neither has the field of paleontology.

Two years ago, Schweitzer gazed through a microscope in her laboratory at North Carolina State University and saw lifelike tissue that had no business inhabiting a fossilized dinosaur skeleton: fibrous matrix, stretchy like a wet scab on human skin; what appeared to be supple bone cells, their three-dimensional shapes intact; and translucent blood vessels that looked as if they could have come straight from an ostrich at the zoo.

By all the rules of paleontology, such traces of life should have long since drained from the bones. It's a matter of faith among scientists that soft tissue can survive at most for a few tens of thousands of years, not the 65 million since T. rex walked what's now the Hell Creek Formation in Montana. But Schweitzer tends to ignore such dogma. She just looks and wonders, pokes and prods, following her scientific curiosity. That has allowed her to see things other paleontologists have missed—and potentially to shatter fundamental assumptions about how much we can learn from the past. If biological tissue can last through the fossilization process, it could open a window through time, showing not just how extinct animals evolved but how they lived each day. "Fossils have richer stories to tell—about the lub-dub of dinosaur life—than we have been willing to listen to," says Robert T. Bakker, curator of paleontology at the Houston Museum of Natural Science. "This is one spectacular proof of that."

At the same time, the contents of those T. rex bones have also electrified some creationists, who interpret Schweitzer's findings as evidence that Earth is not nearly as old as scientists claim. "I invite the reader to step back and contemplate the obvious," wrote Carl Wieland on the Answers in Genesis Web site last year. "This discovery gives immensely powerful support to the proposition that dinosaur fossils are not millions of years old at all, but were mostly fossilized under catastrophic conditions a few thousand years ago at most."

Rhetoric like this has put Schweitzer at the center of a raging cultural controversy, because she is not just a pioneering paleontologist but also an evangelical Christian. That fact alone has prompted some prominent paleontologists to be even more skeptical about her scientific research. Some creationists have questioned her work from the other direction, pressing her to refute Darwinian evolution. But in her religious life, Schweitzer is no more of an ideologue than she is in her scientific career. In both realms, she operates with a simple but powerful consistency: The best way to understand the glory of the world is to open your eyes and take an honest look at what is out there.

Reticent by nature, Schweitzer rarely grants interviews and shies away from making grand pronouncements about her scientific research or her religious faith. Instead of news stories about her stunning findings, she has adorned her office wall with a verse from the book of Jeremiah: "For I know the plans I have for you, declares the Lord, plans to prosper you and not to harm you, plans to give you hope and a future."

SCHWEITZER'S UNCONVENTIONAL VIEW of the fossilized past is rooted in her enduring sense of wonder. When she was 5, her older brother gave her a copy of Oliver Butterworth's *The Enormous Egg*, a fantasy that plays off the then-controversial notion of a close kinship between dinosaurs and birds. She became a dinosaur buff, but as so often happens, with adulthood her interests drifted in other directions. She spent summers selling snow cones and fireworks. She worked with deaf children. She earned an undergraduate degree in communicative disorders and a certificate in secondary education.

In 1989, while dividing her time between substitute teaching and her three children, Schweitzer steered back toward her childhood fascination with dinosaurs. She approached Jack Horner, a renowned dinosaur scientist, and asked if she could audit his vertebrate paleontology course at Montana State University. He appreciated her refreshingly nontraditional mind. "She really wasn't much of a scientist—which is good," says Horner, curator of paleontology at the Museum of the Rockies. "Scientists all get to thinking alike, and it's good to bring people in from different disciplines. They ask questions very differently."

Schweitzer's first forays into paleontology were "a total hook," she says. Not only was she fascinated by the science, but to her, digging into ancient strata seemed like reading the history of God's handiwork. Schweitzer worships at two churches—an evangelical church in Montana and a nondenominational one when she is back home in North Carolina—and when she talks about her faith, her bristly demeanor falls away. "God is so multidimensional," she says. "I see a sense of humor. I see His compassion in the world around me. It makes me curious, because the creator is revealed in the creation." Unlike many creationists, she finds the notion of a world evolving over billions of years theologically exhilarating: "That makes God a lot bigger than thinking of Him as a magician that pulled everything out in one fell swoop."

Schweitzer's career began just as paleontologists started framing their own questions in more multidimensional ways. Until the 1980s, researchers were more likely to be trained in earth science than in biology. They often treated fossils as geologic specimens—mineral structures whose main value lay in showing the skeletal shapes of pre-historic animals. A younger generation of paleontologists, in contrast, has focused on reconstructing intimate details like growth rates and behaviors using modern techniques normally associated with the study of living organisms. "It's taking dinosaurs from being curious fossils to being biological entities," says Hans-Dieter Sues, associate director for research and collections at the Smithsonian's National Museum of Natural History in Washington, D.C.

This shifting perspective clicked with Schweitzer's intuitions that dinosaur remains were more than chunks of stone. Once, when she was working with a T. rex skeleton harvested from Hell Creek, she noticed that the fossil exuded a distinctly organic odor. "It smelled just like one of the cadavers we had in the lab who had been treated with chemotherapy before he died," she says. Given the conventional wisdom that such fossils were made up entirely of minerals, Schweitzer was anxious when mentioning this to Horner. "But he said, 'Oh, yeah, all Hell Creek bones smell,'" she says. To most old-line paleontologists, the smell of death didn't even register. To Schweitzer, it meant that traces of life might still cling to those bones.

She had already seen signs of exceptional preservation in the early 1990s, while she was studying the technical aspects of adhering fossil slices to microscope slides. One day a collaborator brought a T. rex slide to a conference and showed it to a pathologist, who examined it under a microscope. "The guy looked at it and said, 'Do you realize you've got red blood cells in that bone?'" Schweitzer remembers. "My colleague brought it back and showed me, and I just got goose bumps, because everyone knows these things don't last for 65 million years."

When Schweitzer showed Horner the slide, she recalls, "Jack said, 'Prove to me they're not red blood cells.' That was what I got my Ph.D. doing." She first ruled out contaminants and mineral structures. Then she analyzed the putative cells using a half-dozen techniques involving chemical analysis and immunology. In one test, a colleague injected rats with the dinosaur fossil extract; the rodents produced antibodies that responded to turkey and rabbit hemoglobins. All the data supported the conclusion that the T. rex fossil contained fragments of hemoglobin molecules. "The most likely source of these proteins is the once-living cells of the dinosaur," she wrote in a 1997 paper.

That article, published in *Proceedings of the National Academy of Sciences*, sparked a small flurry of headlines. Horner and others regarded Schweitzer's research as carefully performed and credible. Nonetheless, says Horner, "most people were very skeptical. Frequently in our field people come up with new ideas, and opponents say, 'I just don't believe it.' She was having a hard time publishing in journals."

Schweitzer was also stymied by her unconventional fusion of paleontology and molecular biology. "Those are two disciplines we don't usually see in the same sentence," says Lawrence Witmer, an Ohio University anatomy professor. Techniques that were routine in one discipline seemed odd when applied to the other. "If she was working with modern animals, there wouldn't be anything special about what she was doing," says Horner. But molecular paleontology was unheard-of. "It is a wide-open field that she invented," Horner says.

Soldiering on with minimal funding, Schweitzer continued to hunt for the retention of living tissue longer than scientific theory might predict. When a group of fossil hunters found a cluster of preserved bird eggs in a city dump in Neuquén, Argentina, they originally believed the shells contained nothing but sand. Schweitzer placed the remains under scanning electron and atomic force microscopes and concluded that the 70-million-year-old eggs still held embryos containing intact collagen.

For eight years, Schweitzer's career bobbed along with innovative but not attention-grabbing projects. Then she found that stretchy stuff inside a T. rex femur.

SCHWEITZER'S BREAKTHROUGH, like her early insight into the cadaverous odor of dinosaur bones, emerged from the fossil fields of the Hell Creek Formation, rugged badlands so remote that much of it lacks even unpaved roads. Tucked into Montana's northeast corner, Hell Creek was one of the last places on Earth dominated by dinosaurs before they became extinct.

Horner's goal was to conduct a complete census of Hell Creek's dinosaur population—"just go out and collect everything," he says. In 2000, near one of his satellite camps, field crew chief Bob Harmon was eating lunch when he noticed a T. rex foot bone protruding from a sandstone cliff above his reach. Climbing a folding chair balanced on a pile of rocks, Harmon found another bone, then another, then another.

By the time the team had excavated all the bones and encased them in plaster, the collection weighed 3,000 pounds, heavier than the helicopter could lift. With no other way to transport it, scientists reluctantly split the plaster jacket and broke the T. rex's 3.5-foot-long femur. In the process, the fossil bone shed some fragments. Workers wrapped them in aluminum foil and shipped them to North Carolina State University, where Schweitzer had just started teaching. "Jack just gave me the chunks and said, 'See what you can do with them,'" she recalls. Schweitzer, coping with culture shock and a recent divorce, had hit a lull in her research. "I wasn't out there soliciting new projects," she says. "I was trying to survive through each day."

Her lab was still stacked with unpacked cartons when she opened the cardboard box from the T. rex dig and pulled out the biggest fragment. Looking at it with the eyes of a biologist, she immediately saw it was more than a fossil. Time and history began to unwind. "Oh, my gosh," she said to her laboratory assistant, Jennifer Wittmeyer. "It's a girl. And it's pregnant."

What Schweitzer saw was medullary bone, a type of tissue that grows inside the long bones of female birds. Medullary bone is produced during ovulation as a way of storing the calcium needed for egg production; then it disappears. "I looked at it under the dissecting scope," Schweitzer says. "There was nothing else it could be." The medullary bone even contained gaps and mazelike fiber patterns resembling those of modern birds.

Until that moment, no one had ever identified that tissue in a dinosaur, making it impossible to definitively sex such an animal. "Everything we've ever tried to do has been an utter guess," Schweitzer says. For instance, researchers had tried to distinguish a male from a female based on the shape of a creature's body or the size of its head crest. Now they had a way to link gender with morphology and, drawing on parallels with living animals, even with behavior.

The second surprise hit in January 2004. While Schweitzer was attending a departmental taco party, Wittmeyer raced breathlessly into the room. "You aren't going to believe what happened," the lab assistant sputtered.

Wittmeyer had been pulling the late shift, analyzing pieces from the T. rex limb. She had just soaked a fragment of medullary bone in dilute acid to remove some calcium phosphate. This was an unusual procedure to carry out in a dinosaur lab. Scientists typically assume that a fossilized dinosaur consists of rock that would entirely dissolve in acid, but Schweitzer wanted to get a closer look at the fossil's fine structure and compare it with that of modern birds. That night Wittmeyer marveled at a small section of decalcified thighbone: "When you wiggled it, it kind of floated in the breeze."

Schweitzer and Wittmeyer pondered the meaning of the stretchy sample, feeling mystified and ecstatic. The remains seemed like soft tissue—specifically matrix, the organic part of bone, which consists primarily of collagen. Yet this seemed impossible, according to the prevailing understanding. "Everyone knows how soft tissues degrade," Schweitzer says. "If you take a blood sample and you stick it on a shelf, you have nothing recognizable in about a week. So why would there be anything left in dinosaurs?"

Next Schweitzer examined a piece of the dinosaur's cortical bone. "We stuck the bone in the same kind of solution," she says. "The bone mineral dissolved away, and it left these transparent blood vessels. I took one look, and I just said: 'Uh-uh. This isn't happening. This is just not happening.'" She started applying the same treatment to bone fragments from another dinosaur that she had acquired for her dissertation. "Sure enough," she says, "vessels all over the place."

Less than a month later, while Schweitzer was still collecting data on the soft tissue, came a third score. Wittmeyer walked into the lab looking anxious. "I think maybe some of our stuff's gotten contaminated, because I see these things floating around, and they look like bugs," she said. Worried that she would lose her dinosaur blood vessels before she could publish an article about them, Schweitzer rushed to rescue the sample. What she found startled her. Through the microscope she could see what looked like perfectly formed osteocytes, the cells inside bone.

The past was roaring to life.

SCHWEITZER PUBLISHED HER FINDINGS in reverse order—soft tissue first, then the medullary bone—in the journal *Science* last year. The ensuing avalanche of publicity, sometimes couched in breathless hyperbole ("Jurassic Park-type find could be first step in re-creating T. rex," huffed a story in the *Ottawa Citizen*), made her squeamish. She tried to ignore the media, but to no avail. Since the articles appeared, she has become one of the world's best-known paleontologists. Her findings challenge such basic assumptions about animal preservation that her colleagues have put her research—and the woman herself—under the microscope.

If soft tissue can last 65 million years, Horner says, "there may be a lot of things out there that we've missed because of our assumption of how preservation works." James Farlow, a paleontologist at Indiana University–Purdue University at Fort Wayne, adds, "If you can preserve soft tissue under these circumstances, all bets are off."

Schweitzer's work opens the possibility of comparing dinosaur tissue with the tissue of living animals. It could also allow scientists to reconstruct ancient biology, such as prehistoric disease. If paleontologists encounter vascular channels in dinosaur fossils, they might also find nematodes, or roundworms, that lived off the animals' internal organs. "I'll bet you a six-pack of Coors that pretty soon people will be discovering Cretaceous parasites inside Cretaceous bones," says Bakker. "The possibility of looking into epidemiology and pathology is pretty cool."

On the flip side, Jeffrey Bada, an organic geochemist at the Scripps Institution of Oceanography in San Diego, cannot imagine soft tissue surviving millions of years. He says the cellular material Schweitzer found must be contamination from outside sources. Even if the T. rex had died in a colder, drier climate than Hell Creek, environmental radiation would have degraded its body, Bada says: "Bones absorb uranium and thorium like crazy. You've got an internal dose that will wipe out biomolecules."

Others question Schweitzer's thoroughness. "The pictures were stunning, but the paper fell quite short," says Hendrik Poinar, a molecular evolutionary geneticist at McMaster University in Ontario. Schweitzer has not proved that the elastic tissue she found actually consists of molecules from the original dinosaur. Poinar ticks off a list of tests Schweitzer could have conducted, including searching for the building blocks of proteins and then sequencing them to determine their origin. "I understand you want to get your papers out quick and flashy," Poinar says, "but I'm more in favor of longer work with slam-dunk authenticity."

Schweitzer agrees. "I am a slam-dunk scientist," she says. "I would have much rather held the paper back until we had reams and reams of data." But without publishing a journal article, she says, she could never have hoped for funding. "Without the papers in *Science*, I didn't stand a chance," she says. "That's the saddest part about doing science in America: You are totally driven by what gets you funding." Since publishing, Schweitzer has conducted many of the analyses Poinar suggests, with initially promising results.

For a scientist, the ultimate test is having independent researchers replicate your results. So far, there hasn't been a mad rush to do so—few have expertise in both molecular biology and paleontology, not to mention the passion needed to carry out such work. But there is activity. Patrick Orr at University College Dublin is bringing together geologists and organic geochemists to look for soft tissue in a 10-million-year-old frog fossil. Paleontologists at the University of Chicago are setting up a laboratory to look for similar tissue in more T. rex remains; Horner is starting to decalcify other dinosaur bones. In the dinosaur lab at the Children's Museum of Indianapolis, Bakker has taken some peeks. "I haven't found anything yet," he says, "but wouldn't be a bit surprised if soon somebody comes up with more sticky, bouncy stuff."

WHILE SCIENTISTS STRUGGLED to make sense of the bones, another community had no doubt about how to interpret the results. The reports were quickly embraced by biblical literalists who believe God created life on Earth less than 10,000 years ago. For decades they have been working to place a scientific patina on their ideas. The Institute for Creation Research runs a graduate school near San Diego with 11 instructors who hold doctorates in biochemistry, geology, and other sciences. Conferences offer papers on topics like the physics of the Genesis flood. "Any time there's empirical evidence, that's gold for them," says Ronald Numbers, a professor of the history of science and medicine at the University of Wisconsin at Madison.

To Schweitzer, trying to prove your religious beliefs through empirical evidence is absurd, if not sacrilegious. "If God is who He says He is, He doesn't need us to twist and contort scientific data," she says. "The thing that's most important to God is our faith. Therefore, He's not going to allow Himself to be proven by scientific methodologies."

Some creationists, noting Schweitzer's evangelical faith, have tried to pressure her into siding with them. "It is high time that the 'Scientific' community comes clean: meaning that the public is going to hold them ACCOUNTABLE when they find out that they have been misled," reads a recent e-mail message Schweitzer received. She has received dozens of similar notes, a few of them outright menacing.

These religious attacks wound her far more than the scientific ones. "It rips my guts out," she says. "These people are claiming to represent the Christ that I love. They're not doing a very good job. It's no wonder that a lot of my colleagues are atheists." She told one zealot, "You know, if the only picture of Christ I had was your attitude towards me, I'd run."

Ironically, the insides of Cretaceous-era dinosaur bones have only deepened Schweitzer's faith. "My God has gotten so much bigger since I've been a scientist," she says. "He doesn't stay in my boxes."

SCHWEITZER'S RESEARCH DOESN'T STAY within familiar boundaries either. Now there is no clear limit to how far science can go in bringing back the past. In particular, the letters DNA are never far from anyone's lips. "If there's preservation of cells, maybe there's preservation of the constituents of the cells," anatomist Lawrence Witmer says. "It could allow some of the molecular and genetic studies done on modern animals to be potentially used on dinosaur samples." Although scientists consider DNA unstable, in 2003 Schweitzer published a paper outlining several proposed ways the molecule might be preserved. For example, the degradation process itself might produce complex polymers that slow the DNA's further destruction.

At the mention of DNA, minds race to science fiction depictions of cloned dinosaurs. In 2005 a Scottish newspaper announced that, thanks to Schweitzer's work, "scientists are a step closer to . . . bringing the most savage predator ever to walk the earth back from extinction." Even the National Science Foundation blurred the line. When it awarded Horner a grant to study T. rex blood cells years ago, the agency timed the announcement to coincide with the theatrical release of *Jurassic Park.*

Schweitzer scoffs at visions of dinosaur parks. If anyone ever finds dinosaur DNA, she says, it will be fragmented and incomplete. In the unlikely event that scientists could reconstruct a complete dinosaur genome, she doubts that any modern animal could produce an egg capable of growing a dinosaur embryo. And even if that hurdle could be crossed, a viable dinosaur might not last long in 2006: "As far as we know, the way the lung tissue functioned, the way the hemoglobin functioned, was designed for an atmosphere that's very different than today's."

Truth is, Schweitzer hasn't even bothered to look for DNA. She has simply hunkered down to work in her characteristic way: keeping her eyes and her attitude wide open. "So many things are coming together that suggest preservation is far better than we've ever given it credit for," she says. "I think it's stupid to say, 'You're never going to get DNA out of dinosaur bone, you're never going to get proteins out of dinosaur bone, you're never going to do this, you're never going to do that.' As a scientist, I don't think you should ever use the word never."

PATRICIA GADSBY

Cooking for Eggheads

FROM *DISCOVER*

A new breed of chefs is enlisting science in the quest to come up with new recipes and new tastes. Patricia Gadsby tracks down the scientist behind this culinary revolution.

PARIS IS SWELTERING, freakishly hot for an early June morning, and like much of the old city, the lab occupied by Hervé This at the Collège de France, a stone's throw from the venerable Sorbonne, has no air-conditioning. As usual, however, This—pronounced "tiss"—looks dapper in a black suit and one of the impeccable white collarless shirts that have become his trademark. A full day lies ahead in his lab, he says, but first we must shop. He bounds to his feet, ditches his jacket, and descends to the stifling street below, proceeds down a cobbled alley, crosses the boulevard Saint-Michel, rounds a corner, and dives into the local *supermarché*.

He emerges with two dozen eggs and a cold brick of Normandy butter, his face crinkling into a grin. "For our experiments!" he announces. He has yet to break a sweat.

This is head of the molecular gastronomy group in the Collège de France's Laboratory for the Chemistry of Molecular Interactions. That's a mouthful to describe a lab that studies something simple: how the process of cooking changes the structure and taste of food. Nonetheless, molecular gastronomy marks the cutting edge of epicurism these days. Anyone who wields a saucepan is doing chemistry and physics, yet how many of us actually know what's going on in there? Molecular gastronomy aims to apply the piercing clarity of science to the culinary arts. Already in France, which takes the pleasures of the table seriously, molecular gastronomy is an officially recognized, government-funded science.

"Why molecular gastronomy?" asks This, heading off a question he's been asked many times before. "It sounds a little pompous, no? Why not . . . molecular cooking?" Easy, he replies. Cooking aims to produce a dish; it is a craft, a technique. Gastronomy is knowledge, albeit knowledge that can improve your cooking and your appreciation of it. Gastronomy is the science of anything to do with human nourishment, says This, more or less quoting Jean-Anthelme Brillat-Savarin, France's great food philosopher. Writing in 1825, Brillat-Savarin envisaged a discipline that would meld the physics and chemistry of food and cookery with the physiology of eating and especially with the glorious, sensual world of taste.

The "molecular" preface was added in the late 1980s by This and his late colleague, Nicholas Kurti, to evoke the chemical units that make up the water, fats, carbohydrates, proteins, and other compounds in food. Molecular had a dynamic, modern ring to it, perfect for ushering gastronomy into a new era. Besides, molecular gastronomy sounds so much more fun, sophisticated, and cultured than plain old "food science," a field with which it somewhat overlaps but that is largely geared to the mass-market needs of the food industry.

Not that This is a patronizing food snob. (Snobbism would be incompatible with his quest for objectivity, after all.) He would wholeheartedly agree with Brillat-Savarin that "a humble boiled egg" is as worthy of attention as "the banquets of kings." "If all you have to eat is this," he says, plucking an egg from its box and holding it between his thumb and forefinger, "it's important to cook it well."

Do we, though? The standard way to hard-boil eggs in Europe and America—10 minutes in boiling water—is not ideal, says This. The trouble, he notes clinically, is that 212 degrees Fahrenheit is far higher than the temperature at which the egg whites and the yolks coagulate. Egg whites are made up of protein and water (yolks contain fat as well). As eggs cook, their balled-up proteins uncoil into strands, and the strands bind together to form an intricate mesh that traps water. In essence, the proteins form a gel, a liquid dispersed in a solid. Boiling causes too many egg proteins to bind and form dense meshes, "so there is less sensation of water in the mouth," says This. Voilà: rubbery egg whites and sandy, grayish yolks.

The 10-minute egg is just the start of kitchen dogma. Our cookbooks are full of tips, caveats, and stipulations—*précisions*, as This calls them—drawn untested from tradition and folklore. "Cook meat at high temperature to seal in the juices? We've done the test—it's not true," says This. Use only eggs at room temperature for making mayonnaise? Not true either. Season steak with salt before cooking, or salt it afterward? Makes no difference, as the salt doesn't penetrate the meat. Parsing French recipes for a quarter century in his quest for gastronomic clarity, This has identified more than 25,000 such admonitions; so far only a few hundred have been investigated. So many *précisions*, so little time.

THIS, A PHYSICAL CHEMIST and a former editor of the magazine *Pour La Science*, first began his testing as a sideline, alone in a laboratory he'd set up at home. Then he met Kurti, the man who would become his colleague and friend. Kurti was a low-temperature physicist at Oxford University and an irrepressible bon vivant. If there is a father of molecular gastronomy, Kurti is he. Thirty-five years ago, he was already poking the probe of a thermocouple into a cheese soufflé to take its internal temperature, the better to track its vapor-assisted ascent. "We know better the temperature inside the stars than inside a soufflé," Kurti once lamented.

They must have made an odd couple: the short, rotund, Hungarian-born Kurti and the tall, dashing, much younger This. Together they formed the International Workshops on Molecular Gastronomy and began corralling colleagues keen on kitchen science: the American food scholar Harold McGee and the British physicist Peter Barham along with open-minded chefs, critics, and writers who were passionate about food and good-humored enough to put their dearly held ideas (not to mention their egos) to pitiless scientific test. As a meeting place they chose Erice, a monastic town on a Sicilian mountaintop that was already a favorite retreat for physicists like Kurti. Although Kurti died in 1998, the motley group continues to meet every few years to trade information, ideas, and occasional insults, share a few late-night glasses of the local marsala, improvise a test kitchen in a monastery courtyard, and form the foundations for a truly modern cuisine.

The workshops are dizzying affairs. Topics for the texture workshop five years ago included the biomechanics of chewing and swallowing; the structure of meat and how cooking affects it; foams and gels, featuring custards and chocolate mousses; the effects of microwaves on spongy foodstuffs like eggplants and mushrooms; and, on the last afternoon, a marathon session on the fractal nature of baba au rhum dough, conducted by a group of cantankerous physicists. "Well, that's why they are workshops, not lectures—to encourage the free exchange of ideas," This commented at the end of the day.

One participant at the texture workshop was Heston Blumenthal, a radical young British chef and the owner of the Fat Duck, near Windsor. Blumenthal was already receiving raves for the melting tenderness he coaxes out of lamb, achieved through his understanding of how heat diffuses in meat, and for the creation of a fabulous cookie—one that fizzes carbon dioxide in your mouth like so many tiny champagne bubbles. Today the Fat Duck has three Michelin stars and a biochemistry grad student in its development kitchen. Blumenthal is increasingly mentioned in the same breath as Ferran Adrià, the legendary Catalan chef at El Bulli, in Roses, two hours from Barcelona, whose superinventive and rather cerebral cuisine has drawn inspiration from the laboratory for years.

The ascent of the nerdy chef in Europe hasn't gone unnoticed in the United States. Suddenly science—once regarded with suspicion by foodies—looks like the next new thing. The term molecular gastronomy has begun popping up in restaurant reviews and on the food blog eGullet as a label for any edgy, out-there cuisine that combines unusual ingredients and employs techie gadgets.

"Because the phrase was around and catchy, it got applied to anyone experimenting with food," says McGee, an Erice regular. "Let's just say that many people aren't using molecular gastronomy to mean what Hervé means by it."

THE CONFOUNDING OF MOLECULAR GASTRONOMY with a sort of hipster cuisine drives even a patient man like This a little crazy. *Non, non, non*: Molecular gastronomy isn't a cooking style, he insists. "We shouldn't confuse science with technology. Molecular gastronomy is only the science part. It asks: How does something work? What is the mechanism? The application of that knowledge is the cooking part, and that's technology. Cooking is a technique"—his voice softens—"combined with art." He adds, "Here in the lab, we do the science part—experiments."

He introduces a Spanish student whose doctoral thesis investigates the effect of heat on two vegetable pigments, chlorophyll and carotenoids. Other students, using chromatography and nuclear magnetic resonance spectroscopy, for instance, are studying complex mixtures like meat and vegetable stocks. "You will see, we will do some experiments too," This says. "They will be simple, don't worry." You really can try this at home.

We begin by tackling the "standard model," the 10-minute egg. Can it be improved upon? Well, says This, if your grandmother cooked eggs that way for you, and you adored her and her cooking, there'll be no persuading you of a better way. (As This is fond of saying, "The most important ingredient in cooking is love.") But if you're willing to learn a little egg-protein chemistry, you can calibrate your eggs with astonishing exactitude.

Recall that when an egg cooks, its proteins first unwind and then link to form a rigidifying mesh. But not all its proteins solidify at the same temperature. Ovotransferrin, the first of the egg-white proteins to uncoil, begins to set at around 61 degrees Celsius, or 142°F. Ovalbumin, the most abundant egg-white protein, coagulates at 184°F. Yolk proteins generally fall in between, with most starting to solidify when they approach 158°F. Thus, cooking an egg at 158°F or so should achieve both a firmed-up yolk and still-tender whites, since at that low temperature only some of the egg-white proteins will have coagulated.

"Cooking eggs is really a question of temperature, not time," says This. To make the point, he switches on a small oven, sets the thermostat at 65°C, or 149°F, takes four eggs straight from the box, and unceremoniously places them inside. "I use an oven in the lab; it's easier. But if the oven in your kitchen is not accurate, cook eggs in plenty of water, using a good thermometer." About an hour later—timing isn't critical, and the eggs can stay in the oven for hours or even overnight—he retrieves the first egg and carefully shells it. "The 65-degree egg!" he announces. The egg is unlike any I've eaten. The white is as delicately set and smooth as custard, and the yolk is still orange and soft. It's not hard to see why *l'oeuf à soixante-cinq degrés* is becoming the rage with chefs in France. (Salmonella can't survive more than a few minutes at 60°C, or 140°F, so a 65-degree egg cooked for an hour should be quite safe.)

Next, This turns up the oven thermostat to 67°C, or 153°F, and after waiting a while for the eggs inside to reach that temperature—again, he's casual about the timing—he retrieves a second one: "The 67-degree egg!" At this temperature the yolk has just started thickening up—some of its proteins have coagulated, but the majority have not. "Look, you can mold it," he says, scooping out the yolk and manipulating the pliable orangey-yellow ball like fresh Play-Doh. He tries to mold a heart, then settles for a cube.

"Try one," he says, taking a third egg from the oven for me to play with before turning up the heat to 158°F (70°C). The 70-degree egg, when it is finally done, has a moistly set yolk and a very tender white. "So you see, you can adjust the temperature depending on what you want," says This. If you prefer a firmer egg, cook it at 167°F or 176°F. Bear in mind, though, that the most copious of the egg-white proteins sets at 184°F—hence the rubbery results of the 212-degree bath.

So familiar is This with this process that he can tell at a glance the temperature at which an egg was cooked. During lunch at a local bistro, I notice a 65-degree-Celsius egg on the menu, served on a fricassée de girolles. As the plate is set down, This says: "That's not a 65-degree egg. It's a 64-degree egg." The yolk is soft, and the egg white, while completely opaque, is so delicately jelled and fragile that it breaks apart slightly when it is plated. "*Eh, oui*," the chef sighs; he is having *des ennuis* regulating the heat of his stove. Never mind that the presentation isn't completely perfect—the egg, mixing in with the earthy mushroom stew, is delicious.

BACK IN HIS LABORATORY, This puts on a lab coat to protect his shirt "so my wife won't complain that I make spots on it." He breaks a raw egg one-handed, plops the white into a bowl, and starts rapidly whisking. Whisking, of course, incorporates air into the aqueous white. It also causes some proteins in the egg white to unfold. The resulting protein strands then form a mesh around the air bubbles, stabilizing the foam. Usually an egg white produces about half a pint of foam.

Why not more? asks This, whisking like a demon. It can't be lack of air—there's an endless supply—so it must be lack of water. He adds a squirt to the beaten egg white, whisks again, squirts water, whisks some more. The snowy mass keeps growing. "If I went on beating I could get liters and liters—gallons of foam!—from one egg white," he says, pausing at last to wipe his brow. With a thunderstorm brewing, the lab feels muggier than ever, and even This shows signs of wanting a break. "So you see, you only need one egg to make a lot of mousse, enough for a dinner party." However, the foam will be less stable, because the viscosity of that single egg white has been diluted. Whisking hard helps, as smaller bubbles are more stable. So does beating in sugar (or gum), which stabilizes the foam by increasing its viscosity.

The voluminous-egg-white stunt was first used for an educational project in French schools. Later, it occurred to This that it could be put to culinary use. Eight years ago he struck up a collaboration with one of France's most lionized chefs, Pierre Gagnaire. The two regularly rendezvous to brainstorm: This tosses up new culinary concepts based on his scientific musings, and Gagnaire transforms them into elegant recipes. This proposed replacing water in the expanded egg-white foam with a flavorful liquid to make an ethereal perfumed meringue, an invention he named *cristaux de vent*, or "wind crystals." Gagnaire's creation: a soft black olive buried inside a crisp meringue made light as air with the olive's own pickling brine.

This and Kurti envisaged a day when molecular gastronomy would help people cook in entirely different ways; they never guessed that day might come so soon. Quite a few chefs are toying with chemicals pulled from the lab shelf, using new jelling agents, for instance, to encapsulate sauces and liquids in a fragile, jellified skin, like salmon eggs. "It's truly the *terra incognita* of cooking," This says: full of potential for brilliant, thrilling innovation, or for dreadful mischief if explored without discernment. And it's bound to stir up a stew of admiration, bemusement, wonder, and ridicule—in short, gloriously animated debate.

THE FOLLOWING DAY This has an appointment with Gagnaire at his restaurant on a corner of the rue Balzac. You could call their collaboration a tech transfer. "Mad," says This affectionately of his collaborator in the taxi on the way there. "Even madder than me."

Gagnaire is a little late, fresh off the Eurostar from a conference in London. He has a world-weary mien, or maybe he's just tired: intense pale blue eyes, an angular nose, a slight beard, flopping hair—a corsair in chef's whites. The two men embrace happily on meeting and quickly settle down to business in the salon adjoining the dining room. This's laptop comes out, and notebooks appear on the table. This begins serving up a generous helping of ideas, his mind racing; Gagnaire keeps up, peppering him with questions, taking notes, riffing on This's suggestions. Yes, yes, the texture of a taste is very important, says Gagnaire during a discussion of various jelling agents like methylcellulose and alginates. A taste can change with a change in the texture or the surface of the food. Which scientist, Gagnaire wants to know, has studied the effect of surfaces on taste? Later, This reviews plans for an upcoming dinner in which each dish will be named for a scientist and evoke his work: *le dessert Einstein, le plat Faraday, la sauce Pasteur.*

"OK, *d'accord*," Gagnaire says gallantly. Jazz plays softly over the restaurant's sound system. There's a clink of cutlery, a murmur of appreciative conversation from the dining room. Fleets of beautiful little dishes begin to arrive. One, a dish of sweet mild scallops and mussels in a pool of consommé, is enlivened by a rust-colored cream of *araignée de mer*—spider crab—that floods the mouth and nose with sea aromas. Beside it sits a salad of meaty tomatoes, with tangled greens and seaweed on a sauce dark as night, and white wind crystals, the ethereal meringues that Gagnaire impregnates with black olives. It's a deceptively simple combination: plush tomatoes, the tang of slippery seaweed, the meringue crisp and sweetish at first bite, then soft and pungent with olive on the inside.

As soon as this dish is whisked away, I want more. Jean-Anthelme Brillat-Savarin, the original gastronome, was on to something. "The creation of a new dish," he wrote in *The Physiology of Taste*, "does more for the happiness of mankind than the discovery of a new star." If Nicholas Kurti is the father of molecular gastronomy, and Hervé This is its son, then Brillat-Savarin surely is its holy ghost.

GREGORY MONE

Hollywood's Science Guru

FROM *POPULAR SCIENCE*

When Hollywood directors need to make the science in their films believable, or even simply plausible, they turn to a former MIT professor named John Underkoffler. As Gregory Mone shows, Underkoffler is more than an adviser: he is also trying to make the depiction of movie scientists as real as the science.

F LASH BACK TO EARLY 2002: John Underkoffler is leaning over the desk in his small office on a Universal Studios production lot. Just over a year ago, he was a professor at the Massachusetts Institute of Technology. Now he's an adviser on *Hulk*, an adaptation of the popular comic-book series. The film's director, Ang Lee, has questions, and it's Underkoffler's job to answer them. Why is the Hulk green? Why don't poisons affect him? Why do bullets bounce off his skin when he's angry? Lee wants credible explanations that use the latest science, and since the movie is already being shot, he wants them quickly. In effect, he wants Underkoffler to reverse-engineer a superhero.

For months, Underkoffler has been scouring the relevant scientific literature on genetics, chemistry, materials science and animal behavior. He has consulted leading experts, even flying to Boston to meet with iconoclastic Harvard University cellular biologist Donald Ingbet. The walls of his office are plastered with DNA printouts, cell diagrams and sketches of fictional nanotech devices, mementos of his search.

Today he makes the breakthrough. He sprints down the hall to Lee's office and makes a triumphant announcement: "I know why the Hulk is bulletproof!" He explains that the Hulk's DNA must have been spliced with that of another creature, a docile, seafloor-dwelling invertebrate that can harden its body tissues when in peril. The Hulk is . . . part sea cucumber.

Fast-forward three years. Underkoffler is sitting in his own office now, a cavernous, computer-filled lab in downtown Los Angeles. With gray-speckled hair and a goatee, Underkoffler's serene exterior masks a mental energy more common to a heavily caffeinated college kid than a 38-year-old consultant. We're watching the DVD of *Hulk*, which came out in 2003. The movie's opening sequence is science delivered in the quick-cut edits common to music videos. It shows David Banner, the Hulk's mad-scientist father, splicing genes from other creatures into his own DNA, changes that he will pass on to his son. Banner pokes a needle into a jellyfish, shocks a sea cucumber, tests the resistance of a genetically reengineered monkey to poisons. The camera pans over sheets of lab notes—for the movie's purposes they're Banner's scribblings, but Underkoffler wrote every word.

He's quick to admit that Banner's sea-cucumber trick isn't technically possible. But it's conceivable, and that's the point. As one of Hollywood's first full-time science advisers, he can't make everything accurate, but he hopes to reduce the gibberish and mitigate people's fear of science. "One of the things I hate hearing is 'You have to keep it simple.' You don't," he insists. He thinks many filmmakers don't give their audiences enough credit. When actors sound like real scientists, the result is a more believable film. "Even if you don't understand it completely, you get a sense," he says. "It's better than mumbo jumbo."

Technically, Underkoffler is an engineer; that's what the degrees from MIT say. But to do this job he must also be a physicist, a molecular biologist and a cloning expert, not to mention futurist, urban planner, script doctor and inventor. He must be able to think like a brilliant mad scientist, talk like a neurologist, and extrapolate modern technological trends 50 years into the future. And he must do all this with one foot grounded in reality and the other in fantasy.

Underkoffler is not the only science advocate in Hollywood, but his wide range of knowledge makes him unique in the business. In the past five years he has worked on at least eight major studio films, including last year's *Aeon Flux*, about an assassin who tries to overthrow the totalitarian ruler of a seemingly utopian city 400 years in the future, and *The Island*, about human clones bred to provide body parts for others. Currently he's assisting on a movie slated for release in 2007 called *I Am Legend*, about the sole human survivor of a massive pandemic. "The interesting thing about John is that he's such a generalist," says Tim Squyres, the editor of *Hulk* and brother of leading Mars scientist Steve Squyres. "In a movie, where we're not trying to solve some biological problem, where we're trying to tell a story, someone who's just aware of a lot of science is far more valuable because he can suggest interesting notions that a specialist wouldn't necessarily get."

MANY HOLLYWOOD STARS HAVE A STORY about how they were "discovered." Underkoffler's differs from the norm. In the spring of 2000, production designer Alex McDowell was visiting MIT's Media Lab, the future-focused design, engineering and science arm of the university. McDowell was researching advanced technologies for *Minority Report*, Steven Spielberg's thriller set in the year 2054. Spielberg didn't just want far-out stuff; he wanted "future reality." But despite the flood of innovations McDowell witnessed in Cambridge that day, the flexible computer screens and wild-looking vehicle designs, he was most impressed with a young engineer named John Underkoffler. They talked about technology, of course, but also music and movies and books. "He's got this incredible brain," McDowell says. "He's very, very adept as a scientist. But he's also got pop-culture sensibilities, and he's very knowledgeable about filmmaking."

Underkoffler had been involved with the Media Lab since its inception in 1985. He'd worked in holography as part of a research team (which was featured in the January 1991 issue of *Popular Science*). When McDowell met him, he was building the prototype of a machine called a gestural-recognition interface. This device interprets the user's gestures as commands, turning a person's hands into a keyboard and mouse. Spielberg loved the idea, and Underkoffler's invention was adapted for *Minority Report*, which came out in 2002. In the film, the interface helps police officers see the future and arrest killers before they can commit their crimes. At police headquarters, Tom Cruise, playing the detective protagonist, moves his hands through the air to sort through future crime scenes.

Underkoffler was gratified that his invention was incorporated into Spielberg's movie, but he was looking for more. He had begun studying film and literature on the side. He'd even co-founded a comedy troupe, the Hoist Point Orchestra, and had contributed to humorous philosophical riffs on such mysteries as disappearing socks. He was, in a way, already moving beyond the lab when McDowell called him up a year later. "John, have you ever thought of working in Hollywood?"

"Sure."

"Good," McDowell replied. "Get on a plane."

UNDERKOFFLER'S POSITION IS HARDLY GLORIFIED—his name sometimes appears after those of the carpenters, electricians and tailors. But Underkoffler exercises a subtle influence. In *Aeon*

Flux, he acted as urban planner, suggesting that the buildings in a self-contained 25th-century city should be made of bamboo and cement, since metals would be too precious to use in construction. On *The Island*, he was brought in to help make the cloning-related dialogue believable. On *Hulk*, he showed the actors how to handle a confocal microscope and even selected the screensavers for their computers (images of the growth processes of cancerous brain cells borrowed from Ang Lee's wife, a molecular biologist). For the *I Am Legend* job, he is studying up on virology and disaster planning.

Underkoffler throws himself into his research, yet he knows he'll have to let some things go. While watching *Minority Report* with him in his lab, I suggest that the transparent video discs used in the movie seem unnecessary, that in 2054 information would most likely be stored digitally. He winces slightly. Although he was hired to develop the gestural interface, he quickly got more involved, writing impromptu lines of dialogue for scientist characters, pointing out inconsistencies, and, yes, noting that discs would be technological relics by 2054. But he was told that they looked good on film.

As a counterexample, he excitedly skips to a scene in which the Cruise character views a holographic home video of his wife and son. When this scene was in the works, Underkoffler recalls, the holographer in him couldn't stay quiet. He'd worked in the field for years. He understood its limitations, even 50 years out. If he had to choose his battles, this was going to be one of them. The image couldn't be a complete, in-the-round holograph because it was shot from a single angle, with a handheld camera. He pointed out that if this were the case, there couldn't be any back to the images. The filmmakers heeded his advice, and the result—grainy, incomplete holograms—looks like real technology, complete with flaws and glitches. Both suggestions were attempts to make the world on the screen more real, to build a more believable future.

It's not just science, but scientists, that Underkoffler wants to portray realistically. His eyes light up when I mention the 1985 comedy *Real Genius*, about a group of brilliant science students. "They absolutely nailed the Caltech/MIT culture!" He pulls the 1980 movie *Altered States* from his DVD collection, slides the disc into his Mac, and clicks forward to a dinner scene in which a group of scientists are chugging from wine bottles, arguing vehemently about the nature of consciousness. This, he says, is science. Not the booze, necessarily, but the energy and enthusiasm. These aren't detached, Spock-like drones in white lab coats. "[Physicist Richard] Feynman and the really good ones are often just the opposite," he says. "They're passionate and sometimes irrational."

UNDERKOFFLER IS INTELLECTUALLY OMNIVOROUS—his library says it all. The ground floor of his lab holds the light reading, such as *300 Years of Gravitation* and a two-volume set entitled *String Theory*, plus entire bookshelves stacked with works on cinema, art and philosophy. Upstairs in the loft are several hundred works of serious literature, including 17 novels by the influential postmodernist author John Barth. The organized engineer in him is also apparent: All the books, along with hundreds of DVDs, are sorted by subject and alphabetized.

Along with his Hollywood accomplishments, Underkoffler remains a working engineer. With funding from defense and aerospace contractor Raytheon, among others, his research team recently completed a working version of the gestural-recognition interface in his L.A. lab. Raytheon thinks it might be useful for military planning, but when I was visiting, Underkoffler presented it to a Hollywood cinematographer as a potential tool for filmmakers.

The success of *Minority Report* in particular has earned Underkoffler acclaim outside Hollywood. He speaks at technology conferences around the world. He advises videogame developers, helping them construct realistic futures. His popularity raises the question of why the science adviser isn't a more common breed. "If you're doing a cop show, you've always got a cop on set. It's the same with doctors," Underkoffler says. "But people rarely get science advisers."

Occasionally a director will bring in a scientist to mine his or her area of expertise—for example, Jet Propulsion Laboratory geophysicist Richard Terrile worked on 2003's journey-to-the-center-of-the-Earth movie *The Core*. And recently, a Harvard mathematician and University of Oxford biochemist teamed up to form a consulting firm, Hollywood Math and Science. So far, their work has been limited to the math-based TV show *Numb3rs*.

McDowell suggests that the problem isn't that filmmakers are averse to science consultants but that good advisers are hard to find. "On the whole, you contact a specialist and they tell you what they know, and then you extract from it what you can," he says. "John is much more proactive. He'll go, 'Well, if that's what you're thinking about, you might be interested in this.'"

Hulk editor Tim Squyres hopes that this man of many talents will someday move from the margins to the center of the action. If Underkoffler wants to cram all his goals into a single film, if he wants scientist characters conveying passion for their subject, if he wants to draw fantasy closer to reality by basing ideas on actual research, then he is going to have to be more than an adviser. "I keep telling John to write a script," Squyres says.

About the Contributors

LAWRENCE K. ALTMAN, M.D., one of the few medical doctors working as a reporter, joined the *New York Times* in 1969. An award-winning journalist, he began his career at the Centers for Disease Control and Prevention, and as chief of the Public Health Service's division of epidemiology and immunization. A graduate of Harvard and Tufts University School of Medicine, Dr. Altman is a professor at New York University School of Medicine, a Master of the American College of Physicians, a Fellow of the American College of Epidemiology and the New York Academy of Medicine, and a member of the Institute of Medicine of the National Academy of Sciences. In addition to reporting, he writes the "Doctor's World" column in Science Times.

"At age 98, Dr. DeBakey continues to be a medical educator," he reports, "this time as a patient who has shared one of the most amazing stories in the annals of medicine. He showed that a strong elderly person can survive a difficult operation. Beyond that, his story has been widely discussed because it illustrates the ambiguity of his advance directives and raises many practical and ethical questions that affect patients in hospitals around the world. Among them:

"Under what circumstances can a family member and an ethics committee override a patient's 'do not resuscitate' order?

"How often do doctors clarify in discussions with a patient the measures that are generally included in the advance directives that they sign when seriously ill?

"How well do seriously ill patients understand such directives?

"How many hospitals that can deliver the kind of care Dr. DeBakey received would give it to someone who was not a local icon or a VIP? And would the hospital pay for a non-VIP's care?"

WILLIAM J. BROAD is a senior writer at the *New York Times* and author or coauthor of seven books, including *Germs: Biological Weapons and America's Secret War*, a number-one *New York Times* bestseller. For more than two decades, he has covered science for the *Times*, reporting on such topics as geology, astronomy, oceanography, biology, physics, ecology, astrophysics, space weapons, and nuclear arms. Mr. Broad has won two Pulitzer Prizes with *Times* colleagues, as well as an Emmy and a duPont. He holds a master's degree in the history of science from the University of Wisconsin.

He writes: "Like many people, I thought that carbon dioxide in the atmosphere was rising to record highs and that a global meltdown would ensue if it rose much higher. Then, one day, I stumbled on a graph that stopped me cold. It showed past CO_2 trends. But the zigs and zags, instead of portraying changes over thousands of years, as climatologists often depict, showed them over millions of years. Contrary to the usual claims, the graph revealed that CO_2 levels had generally dropped over time, so much so that past concentrations were up to eighteen times higher than today's. On this scale, the recent CO_2 rise barely registered. It seemed like an inconsequential blip.

"Curious, I began researching the story behind the graph and found more surprises. It turned out that one group of scientists saw the planet's long history of CO_2 fluctuations as poorly correlated with its cycles of hot and cold, further undermining the claimed supremacy of carbon dioxide. The skeptics did not inhabit the scientific fringe, either. They often worked at top schools and published in such respected journals as *Nature* and *The Proceedings of the National Academy of Sciences*. I found other scientists who argued that the planet's CO_2 swings did predict climate trends fairly well, if not perfectly. Whoever was right, it was clear to me that, on the subject of climate change, the usual pronouncements of scientific consensus hid a rather fascinating debate unknown to the public.

"The article I wrote for the science section of the *New York Times* tried to lay out the discussion as unemotionally as possible, aiming for clarity and balance. It may not be the most absorbing story I have ever written. But in terms of depth and potential importance, I see it as one of my best. At a minimum, the emerging science of climate reconstruction shows that the situation is more complex than Al Gore would have us believe and, at most, might develop to the point that it affects the prevailing view on global warming and the widespread jitters over atmospheric carbon. Because of that, the field seems worth funding robustly and following closely."

STACEY BURLING is a reporter for the *Philadelphia Inquirer*. She has spent most of her years at the newspaper covering various aspects of medicine, including health policy and medical research, quality, and delivery of care. She is currently assigned to the business desk. Before joining the *Inquirer*, she worked at newspapers in Denver; Norfolk, Virginia; and Moline, Illinois. She has covered education, police, and courts, and has worked as a general assignment project reporter. A native of South Bend, Indiana, Burling now lives in the Philadelphia suburbs with her husband, two sons, and three cats.

"The story of Bob Moore's brain began almost two years before his death," she explains. "It grew out of others I had done about the social consequences of Alzheimer's disease. I wanted to write a science story about what dementia does to a brain. The challenge was to make that comprehensible and compelling. The story of one patient's autopsy was a way to combine hard science with biography. My goal was to link what happened in the brain on a cellular level to the behavioral changes that are so devastating for families. I wanted to start with a patient who was still alive so that readers could see more clearly why this science matters. I was lucky to find Bob Moore, who had led an interesting life and had a very open family."

TYLER CABOT is an associate editor at *Esquire*, where he oversees the magazine's Best and Brightest issue, an annual collection of articles honoring innovators and change-makers in science, politics, and the arts, in which "The Theory of Everything" first appeared. He also edits the magazine's literary, travel/adventure, and automotive coverage.

"The first reaction I got from many physicists in my reporting was, '*Esquire*?!? Why would *Esquire* do an article on string theory?' " he recalls. "But it became apparent very soon after I began reporting that there was much more to this story than equations and theories. The physics world is in transition. People are tired of just hypothesizing. They want results. And with the opening of the Large Hadron Collider near Geneva, some long-held theories and hypotheses will finally get tested. So I had the good fortune of catching the physics community at a very critical and climatic moment."

JOHN CASSIDY has been a staff writer at *The New Yorker* since 1995. He has published more than fifty major articles in the magazine, on subjects ranging from the Iraqi oil industry, to George Steinbrenner, to the economics of Hollywood movie making. A number of his articles have been nominated for National Magazine Awards, including "The Greed Cycle," an essay about corporate avarice, which was published in November, 2002, and "The David Kelly Affair," an account of the mysterious death of a British weapons scientist, which appeared in December, 2003.

" 'Mind Games' was long in the making," he explains. "It was back in the nineties that I first thought of writing about behavioral economics, a new school of thinking that challenged many old nostrums. But it wasn't until the rise of neuroeconomics, which arose out of behavioral economics, that I could conceive of a piece that would interest my fellow economists and general readers alike. Researchers using brain scanners to examine how people make decisions about money: it wrote itself, almost."

A great-great-grandson of Charles Darwin, MATTHEW CHAPMAN has written and directed five independent films and written or co-written several others. He is the author of *Trials of the Monkey: An Accidental Memoir*, which was favorably compared to the works of Kingsley Amis, Robert Graves, and Paul Theroux. His second book, *40 Days and 40 Nights: Darwin, Intelligent Design, God, OxyContin and Other Oddities on Trial in Pennsylvania*, is an account of *Kitzmiller v. Dover*, the evolution-versus-intelligent-design trial in Pennsylvania, and the many odd characters involved. He is currently preparing a feature film, *The Ledge*, which he wrote and will direct in 2007.

"Since writing 'God or Gorilla' for *Harper's* magazine, I have become increasingly fascinated by the conflict between science and faith," he comments. "Watching educated advocates of intelligent design contort themselves in order to defend scientifically absurd religious concepts while all the while pretending they were adhering to the principles of science was both disturbing and sad. They were themselves the first victims of this decision to choose faith in God above the obvious truth of evidence, and if they, with all their advantages, felt psychologically compelled to make this choice, how could one not sympathize with—and be increasingly terrified of—less educated fundamentalists of all types? In my view, nothing could be more important for the survival of our species than the study of how we delude ourselves."

JENNIFER COUZIN is a staff writer for *Science* magazine, where she covers various issues in medicine and basic biology. Her work has also appeared in *U.S. News & World Report*, *Newsweek*, and the *Washington Post*, among other publications. In 2004 she won the Evert Clark/Seth Payne Award, given annually to a young science journalist, and a story she wrote about dueling researchers studying aging appeared in *The Best American Science Writing 2005*. She grew up in Toronto and lives with her husband in Washington, D.C.

"This story came to me via a local Wisconsin newspaper," she explains. "An article on the case mentioned that the professor had been turned in by her graduate students. I began to wonder: what must it be like to uncover evidence that the mentor who is shaping your career is breaking a cardinal rule of science?

"We journalists are taught to detach ourselves from the people we write about. With this story, I had to work hard to maintain that emotional distance. I couldn't help but admire these students for their poise and honesty, for how they had handled the data falsification they'd unwittingly discovered, for talking with me for hours about such a difficult time. The story exposed the flip side of scientific misconduct: so often we focus on the researchers who commit fraud, and forget about the young scientists in whose lives they play a starring, even parental, role.

"These days, the students are moving on and putting those difficult months behind them. But their brush with fraud left a lasting impression. In an e-mail she wrote me five months after the story was published, Chantal Ly said she's realized that 'graduate students are at the bottom of the food chain.' She's grateful to have left science behind, and plans to try her hand at business. Jacque Baca, who continues with her PhD program at the University of Wisconsin, says that her experience in the Goodwin lab 'will probably stay with me throughout my career.'"

When he's not tracking down unusual science and technology stories, *Wired* contributing editor JOSHUA DAVIS spends his free time competing in the world's wildest competitions. It began when he entered the U.S. Armwrestling Nationals, where he lost every single match but ended up fourth (out of four) in the lightweight division. That earned him a spot on Team USA and sent him to Poland for the World Armwrestling Championships. *The Underdog*, his book about becoming an internationally ranked armwrestler (as well as a sumo wrestler, matador, and backward runner), is now available from Random House. His writing can be found at www.joshuadavis.net.

"Early in 2006, I bumped into my favorite college professor," he remembers. "I hadn't seen him in years, but we fell into an easy conversation about the new opera season. He had taught a small seminar focused on Mozart and Wagner, and the class had spent a lot of time together watching performances. I felt like we had known each other well and was happy to see him but after a few minutes he apologized and told me he had no idea who I was. He quickly explained that it wasn't my fault—he was born with a neurological condition that prevented him from processing faces. When I said my name, he remembered exactly who I was and the conversation moved on. But the idea that there was a group of people in the world who couldn't see faces captivated me, and I spent the next six months learning everything I could about prosopagnosia."

DAVID DOBBS writes freelance for the *New York Times Magazine, Scientific American, Scientific American Mind, Audubon, Slate, New Scientist,* and other publications. He also keeps a blog on science, medicine, nature, and culture, Smooth Pepples, and edits and moderates Mind Matters, *Scientific American*'s expert-written "blog seminar" on mind and science. Dobbs's *New York Times Magazine* feature about the decline of the autopsy, "Buried Answers," was included last year in *The Best American Science and Nature Writing 2006.* He is the author of three books, most recently *Reef Madness: Charles Darwin, Alexander Agassiz, and the Meaning of Coral* (Pantheon, 2005), which Oliver Sacks found to be "brilliantly written, almost unbearably poignant," and "an enthralling picture of three grand scientific minds." You can find more of Dobbs's writing at his blog, smoothpebbles.com, or his website, daviddobbs.net.

"To me," he writes, "this surgical treatment struck directly at what makes depression so frustrating and compelling. Everybody knows that stress and enviroment can cause or aggravate depression. Yet the depressive experience has always seemed most centrally a battle with something that seems to rise from within. This apparently internal genesis generates many of the various sentiments about depression: the sufferer's embarrassment, as if depression revealed some moral taint rooted in family or character; the attitude among many nonsufferers that depression reflects a passivity that one need only rally to defeat; and the idea—popular among writing types, I'm afraid—that depression expresses something fundamental about a person, and that to remove it would be to diminish the sufferer's self.

"The startling cures experienced by Deanna and other patients proves all such sentiments ill-founded—even while showing that depression really does rise at least partly from within. Cause and cure were internal, but in the way of a weak immune system: a malfunction within that makes one more vulnerable to pathogens without. At the same time, the way these patients connected to the world when the implants flicked on—their sudden re-engagement on the terms most central to their characters—buried any notion that depression is integral to self. Depression didn't express their selves; it smothered them. I still think about these people often."

PATRICIA GADSBY is a contributing editor for *Discover* magazine. In her former guise as a senior editor at *Discover*, she covered biology and medicine and conceived two special issues that were nominated for National Magazine Awards, one of them winning an Ellie. In an earlier guise still, she wrote a cookbook, and now (with a science slant) she's returned her attention to food and eating. She has written about the paradox of the old Inuit diet, the microorganisms in sourdough, the sense of taste, the chemistry of "fishiness," and the genetic history of chocolate. She has a master's degree in modern languages from Cambridge University, England, and lives with her scientist husband in New York and in Woods Hole, Massachusetts.

She comments: "I'd wanted to write about the phenomenon of molecular gastronomy ever since attending an eccentric and fractious workshop in Erice in Sicily. What impressed me was the fervor participants brought to applying science to better their understanding of food and cooking. For this article I picked one voice out of the multilingual cacophony and fixed on the French chemist Hervé This—a star in France, but hardly a household name here—and hewed to his interpretation of molecular gastronomy. (Of any other participants who feel they did not get their due, I beg forbearance.) I also limited my inquiry to eggs, because everyone, supposedly, can boil an egg. A day in the lab with This, and I could see for myself that the key to cooking eggs well was temperature, not time as we're led to believe. Once you 'get' that the various proteins in the whites and yolks of eggs coagulate at a range of different temperatures, you can use temperature to 'set' the egg textures you want. Surprise—that familiar old egg was new again.

"Granted, some people may find This's approach to cooking reductionist. But if you're anything like me then envisaging how an egg's proteins coagulate actually amps the experience of eating it. That's the thing about the scientific study of all things gastronomical. Everybody eats, quite a few of us cook. For non-scientists it can still be something of a participatory sport."

ATUL GAWANDE is a general and endocrine surgeon at Brigham and Women's Hospital in Boston, an assistant professor at Harvard Medical School and the Harvard School of Public Health, and a staff writer for *The New Yorker*. He is the author of *Better* and of *Complications*, which was a National Book Award finalist.

"Researching this essay was disconcerting," he says. "It forced me to think hard about just how transparent I'd be willing to be about my own practice, and about how difficult the reality of the bell curve is for people both inside and outside medicine. I write partly because I believe in the value of demystification. But there is still fierce disagreement in medicine about whether demystification is good for people—and about how performance is actually improved. The result was that I kept poking away at these questions, and the essay led directly to the new book I wrote this year, *Better*."

DENISE GRADY has been a reporter in the science news department of the *New York Times* since September 1998, and has also worked as an editor there. She wrote for the *Times* for several years before that as a freelancer. She has written more than 500 articles about medicine and biology for the *Times*; edited two *Times* books, one on women's health and one on alternative medicine; and has written *Deadly Invaders*, a book about emerging viruses that was published in October, 2006.

She adds this update to her story: "Like many people who have major surgery, Chris Ratuszny suffered some setbacks—transient speech problems, a drug-resistant infection, and severe spasms of blood vessels inside his skull. All subsided gradually, and Mr. Ratuszny emerged intact—very much himself, to the delight of his family. He went home 16 days after the operation—walking, talking, cracking jokes, and eager to see his 3-year-old son, Sam, and return to his job as a Lexus mechanic. Less than three weeks after that, he went back to work. Dr. Langer expected him to make a full recovery, and lead a normal life."

JEROME GROOPMAN holds the Dina and Raphael Recanati Chair of Medicine at the Harvard Medical School and is chief of experimental medicine at the Beth Israel Deaconess Medical Center. He serves on many scientific editorial boards and has published more than 150 scientific articles. His research has focused on the basic mechanisms of cancer and AIDS and has led to the development of successful therapies. His basic laboratory research involves understanding how blood cells grow and communicate ("signal transduction"), and how viruses cause immune deficiency and cancer. Dr. Groopman also has established a large and innovative program in clinical research and clinical care at the Beth Israel Deaconess Medical Center. In 2000, he was elected to the Institute of Medicine of the National Academy of Sciences. He has authored several editorials on policy issues in *The New Republic*, *Washington Post*, *Wall Street Journal*, and *New York Times*. He has published four books, most recently *How Doctors Think* in March 2007.

"A friend who is a surgeon had told me that his hospital recently instituted a program to permit family members to witness a loved one's resuscitation in the ER," he writes. "He was adamantly opposed to this practice. I was intrigued to learn how such a policy came about, not only at his hospital but at others in the country. As I interviewed advocates and opponents of 'family presence,' I was struck by how each camp was impassioned in its beliefs, despite the limited data on the risks and potential benefits of the practice."

ROBIN MARANTZ HENIG has written eight books, most recently *Pandora's Baby: How the First Test Tube Babies Sparked the Reproductive Revolution*, which won Best Book awards from both the American Society of Journalists and Authors and the National Association of Science Writers. She is a contributing writer at the *New York Times Magazine*, and she has also written for *Civilization, Discover, Scientific American*, and just about every women's magazine in the grocery store. She was a finalist for a National Book Critics Circle Award for *The Monk in the Garden: The Lost and Found Genius of Gregor Mendel, the Father of Genetics*. She and her husband, Jeff, a political scientist at Teachers College at Columbia University, have two grown daughters and live in New York City.

"I consider myself an honest person, and I generally expect other people to be honest, too," she remarks. "I joke that I'm not smart enough to keep track of which lies I might have told to whom, so I just tell the truth; it's easier. But writing this article about high-tech lie detection made me think about the social and psychological value of deception. I don't mean only the little white lies that allow us to get along with one another. I mean bigger lies, too, which I now see as a way to keep a tiny part of ourselves completely and inviolably private. I had thought I was transparent to the people who knew me best. But I'm happy to have discovered, after writing this article, that I'm not transparent after all. None of us is."

JONATHON KEATS is a writer and conceptual artist living in San Francisco. He is the author of two novels, as well as a devil's dictionary of technology, and his many scientifically based conceptual artworks, commissioned by institutions including the Judah L. Magnes Museum and the San Francisco Arts Commission, have been featured on PBS, NPR, and the BBC World Service. He writes for publications including *Popular Science*, *Wired*, and the *Washington Post*, and has been awarded Yaddo and MacDowell fellowships for his fiction, most recently a cycle of fables based on Talmudic legend.

"Shortly before I interviewed John Koza," he recalls, "a computer running his genetic programming was the recipient of the first patent ever awarded to a non-human inventor. For him, this was the ultimate proof-of-concept, a sort of bureaucratic Turing test demonstrating that his software was creative. More interesting to me, though, and to many who have read my article, is the potential of his machine to demystify human ingenuity."

ELIZABETH KOLBERT is a staff writer for *The New Yorker* and author of *Field Notes from a Catastrophe: Man, Nature, and Climate Change* (Bloomsbury, 2006). Her three-part series on global warming, which appeared in *The New Yorker* in the spring of 2005, won a National Magazine Award, the National Academies Communications Award, and the American Association for the Advancement of Science's magazine writing award. She lives in Williamstown, Massachusetts, with her husband, John Kleiner, and their three sons.

"Even people who are fairly knowledgeable about global warming tend to think about it as something that will affect the future," she writes. "I want people to realize that significant changes are occurring *right now*. What we have done to the climate is already having a measurable impact on plants and animals around the world. It's quite likely that climate change will wipe out species that we as humans hold dear. Perhaps it will wipe out some that we depend on. Meanwhile, those species most likely to benefit are ones that humans have spent a great deal of energy trying to eradicate; for example, it's been shown that poison ivy thrives in a high CO_2 environment."

GREGORY MONE is a contributing editor for *Popular Science* magazine and the author of *The Wages of Genius* (Carroll & Graf, 2003), a novel about a businessman who thinks he's the reincarnation of Einstein. A 1998 graduate of Harvard College and, later on, New York University's Science, Health and Environmental Reporting Program, Mone also writes for *Technology Review, Women's Health,* and *National Geographic Adventure.* He lives in Canton, Massachusetts, with his wife and baby daughter, and works in their dark, cold basement.

He writes: "For a while I'd been intrigued by the nonsense that passes for science-speak in the movies, so it was fascinating to spend a few days with John Underkoffler, an MIT-trained scientist who consults on Hollywood films, and learn about how things work behind the scenes. In some cases he has the director's ear, and can really effect change, but on other sets he's very much in the background, playing a more subtle role. For the Will Ferrell movie *Stranger Than Fiction,* for example, he wrote a math-focused 30-page treatise on how to think like an obsessive compulsive—just for the actor's background research. Besides spending time with Underkoffler, which was tremendously stimulating, reporting the story was a blast. While preparing to meet with him one morning, for instance, I spotted Bill Nye the Science Guy in a coffee shop in Santa Monica and introduced myself. He had a few salient things to say about science and Hollywood, and after we finished our coffees, we made plans to go surfing together the next morning. We did, and it all felt very L.A."

SYLVIA NASAR is a former *New York Times* correspondent and the author of *A Beautiful Mind.* She teaches at Columbia University's Graduate School of Journalism. DAVID GRUBER is the coauthor of *Aglow in the Dark: The Revolutionary Science of Biofluorescence.* He holds a master's degree in journalism from Columbia and a PhD in marine oceanography Rutgers University. Nasar and Gruber took four months and traveled to Beijing and St. Petersburg to report the tangled tale of the Poincaré conjecture and the reclusive genius who solved it.

OLIVER SACKS is a neurologist in New York City, where he sees patients and teaches at Columbia University. His previous books include *The Man Who Mistook His Wife for a Hat* and *Awakenings*, and his work has been honored by the Guggenheim and Alfred P. Sloan foundations. He is a fellow of the Royal College of Physicians, the American Academy of Arts and Letters, and the American Academy of Arts and Sciences. His most recent book is *Musicophilia: Tales of Music and the Brain.*

"Writing 'Stereo Sue' was a joy for many reasons," he says. "First, I am a card-carrying member of the American Stereoscopic Society and the International Stereoscopic Union, and have always been a passionate stereo buff. Sue Barry, my subject, was a pleasure to write about, since she is so insightful, so lyrical, so thoughtful and articulate; I hope she will one day write her own book on her experiences. And finally, though I so often write about people adapting to losses or deficits of one sort and another, this was a rare chance to write about someone adapting to an unexpected and thoroughly delightful *gain*, the acquisition of an extra sense, so to speak."

BARRY YEOMAN is a freelance journalist who specializes in putting a human face on complex issues, including science, politics, and religion. A winner of numerous national awards, he was named by *Columbia Journalism Review* as one of nine investigative reporters who are "out of the spotlight but on the mark." His writing appears in *Discover*, *Mother Jones*, *AARP The Magazine*, and *O: The Oprah Magazine*. He lives in Durham, North Carolina. His work can be read at barryyeoman.com.

"As a journalist, I like finding opportunities to challenge readers' stereotypes," he explains. "Mary Schweitzer's story was the perfect vehicle for this: Here was an evangelical Christian defending evolution against her coreligionists. As Americans continue to debate the teaching of evolution and creationism in our schools, Schweitzer's work serves as a reminder that the debate cannot be reduced to science versus religion. One does not necessarily exclude the other."

Permissions

A Note from the Series Editor

Submissions for next year's volume can be sent to:

Jesse Cohen
c/o Editor
The Best American Science Writing 2008
HarperCollins Publishers
10 E. 53rd St.
New York, NY 10022

Please include a brief cover letter; manuscripts will not be returned. Submissions can be made electronically and sent to jesseicohen@ netscape.net.

THE BEST AMERICAN
SCIENCE WRITING
2004
Dava Sobel, Editor
Jesse Cohen, Series Editor

ISBN 978-0-06-072640-9 (paperback)

THE BEST AMERICAN
SCIENCE WRITING
2003
Oliver Sacks, Editor
Jesse Cohen, Series Editor

ISBN 978-0-06-093651-8 (paperback)

THE BEST AMERICAN
SCIENCE WRITING
2002
Matt Ridley, Editor
Jesse Cohen, Series Editor

ISBN 978-0-06-093650-1 (paperback)